WEYERHAEUSER ENVIRONMENTAL CLASSICS

William Cronon, Editor

WEYERHAEUSER ENVIRONMENTAL CLASSICS are reprinted editions of key works that explore human relationships with natural environments in all their variety and complexity. Drawn from many disciplines, they examine how natural systems affect human communities, how people affect the environments of which they are a part, and how different cultural conceptions of nature powerfully shape our sense of the world around us. These are books about the environment that continue to offer profound insights about the human place in nature.

The Great Columbia Plain: A Historical Geography, 1805–1910 by D. W. Meinig

Mountain Gloom and Mountain Glory: The Development of the Aesthetics of the Infinite
    by Marjorie Hope Nicolson

Tutira: The Story of a New Zealand Sheep Station by Herbert Guthrie-Smith

A Symbol of Wilderness: Echo Park and the American Conservation Movement
    by Mark Harvey

Man and Nature: Or, Physical Geography as Modified by Human Action
    by George Perkins Marsh; edited by David Lowenthal

Conservation in the Progressive Era: Classic Texts edited by David Stradling

DDT, Silent Spring, and the Rise of Environmentalism: Classic Texts
    edited by Thomas R. Dunlap

Reel Nature: America's Romance with Wildlife on Film by Gregg Mitman

The Environmental Moment, 1968–1972, edited by David Stradling

Weyerhaeuser Environmental Classics is a subseries within Weyerhaeuser Environmental Books, under the general editorship of William Cronon. A complete listing of the series appears at the end of this book.

THE
ENVIRONMENTAL MOMENT

1968–1972

EDITED BY DAVID STRADLING

Foreword by William Cronon

UNIVERSITY OF WASHINGTON PRESS
Seattle and London

The Environmental Moment, 1968–1972 is published with the assistance of a grant from the Weyerhaeuser Environmental Books Endowment, established by the Weyerhaeuser Company Foundation, members of the Weyerhaeuser family, and Janet and Jack Creighton.

© 2012 by the University of Washington Press
Printed and bound in the United States of America
Design by Thomas Eykemans
20 19 18 17   5 4 3 2

UNIVERSITY OF WASHINGTON PRESS
www.washington.edu/uwpress

LIBRARY OF CONGRESS CATALOGING-IN-PUBLICATION DATA
Stradling, David.
The environmental moment, 1968-1972 / David Stradling.
    p. cm. — (Weyerhaeuser environmental classics)
ISBN 978-0-295-99181-8 (pbk. : alk. paper)
1. Environmentalism—United States—History.
2. Environmental policy—United States—History.
3. Environmental protection—United States—History.
1. Title.
GE197.S77 2012   304.20973'09046—dc23   2011041274

The paper used in this publication is acid-free and meets the minimum requirements of American National Standard for Information Sciences—Permanence of Paper for Printed Library Materials, ANSI Z39.48.1984.

For Nina

CONTENTS

Artist Ron Cobb's ecology symbol first appeared in late 1969, and it quickly gained wide usage, from flags and banners to buttons and bumper stickers. The symbol, a creative combination of the letters "E" for environment and "O" for organism, came to represent Earth Day in 1970 and eventually the entire environmental movement.

WILLIAM CRONON

FOREWORD

THE MYRIAD TRIBUTARIES
OF A WATERSHED MOVEMENT

During my sophomore year of high school in Madison, Wisconsin, I learned with considerable excitement in the fall of 1969 that Senator Gaylord Nelson had issued a call for a national "teach-in" to encourage Americans to discuss, learn about, and take action to protect the environment against threats like the pollution and overuse of natural resources that endangered the long-term health of humanity and even the planet itself. Called Earth Day, Nelson's teach-in followed in the wake of widespread popular protests associated with the Civil Rights movement and the Vietnam War. Madison had emerged as a major center for such protests, and even our relatively sheltered suburban high school was drawn into the volatile and sometimes angry activism.

Earth Day undoubtedly gained much of its energy from the already well-organized networks created by other movements, but it also drew on older sources as well: fears of toxicity galvanized by the publication of Rachel Carson's *Silent Spring* in 1962; anxieties emanating from the threat of nuclear annihilation in the Cold War between the United States and the Soviet Union; longstanding concerns about the conservation of natural resources, stretching back to the late nineteenth and early twentieth centuries; and the extraordinary photograph of "Earthrise" snapped by an Apollo 8 astronaut on Christmas Eve in 1968, an image that became an icon for Earth Day in April 1970.

But I remember another inspiration as well: the "Ecology Symbol" designed by the cartoonist Ron Cobb in 1968 and explicated on an 8.5-by-11-inch handout that he released into the public domain in October 1969 so that it could be endlessly photocopied and circulated among environmental activists. I don't remember when I first saw and obtained this remarkable document, but I will never forget it. I still have a copy among the most precious documents of my childhood. (You can explore it for yourself on the page opposite this preface.)

Cobb had sought to create a symbol for environmentalism to emulate the well-known Peace Symbol that had already played such an important role in the anti-war movement. It could hardly have been simpler: a circle inside an ellipse inside a square, with a horizontal line bisecting the circle. Cobb said that these represented the letter "e," standing for "earth" and "environment" inside the letter "o," standing for wholeness and unity. But if you study the handout he created that circulated in so many thousands of copies during the spring of 1970 and thereafter, you'll see that it gestures at many other symbols as well, all implying that the environmental politics it endorsed were tied to far deeper material, ethical, and spiritual traditions reaching back across the history of humanity: Darwinian evolution, modern science, astrology, alchemy, Taoism, Jungian psychology, Buddhism, and many others. The Ecology Symbol is a potent document that on a single page conveys more about the complex impulses giving rise to the environmental movement at the end of the 1960s than many paragraphs of analytical prose could possibly do.

As such, it is also a lovely symbol for what David Stradling has sought to do in this latest addition to the Weyerhaeuser Environmental Classics series, *The Environmental Moment, 1968–1972*. When we first conceived this Classic Texts series nearly a decade ago, the idea was to assemble wide-ranging collections of primary documents relating to important themes in environmental history so students and other interested readers could gain insights that only become possible when we read and understand what people at the time were saying and doing.

Because David Stradling is a brilliant teacher who knows how to make the past come alive in just this way, we recruited him to produce the first volume of Classic Texts, *Conservation in the Progressive Era*. The book succeeded beyond our wildest dreams. By juxtaposing documents that had never before appeared in such close proximity to each other and by using their eclecticism to suggest the complex undercurrents and contradictions that had given rise to progressive conservation at the dawn of the twentieth century, Stradling made this long-ago precursor of modern environmentalism vividly interesting even for readers who might otherwise have found it quite dull.

Although I expected *Conservation in the Progressive Era* to work well for my students when I first assigned it in my American environmental history course, I never expected it to be one of the most successful and popular volumes I have ever shared with students. But that is what it proved to be. Almost no other book I know enables students and other readers to see for themselves how the generalizations that scholars make about the past emerge from the much more complicated sources that serve as partial evidence of those generalizations. Although the book offers crucial insights into conservation politics in the United States, it conveys even deeper truths about how we ask questions and seek answers about any topic as we seek to turn the raw materials of the past into the more seamless prose in which written history typically expresses itself. It's a superb piece of work, and if you've not yet had the pleasure of perusing it, I urge you to do so.

Given how effective Stradling's book on the conservation movement has proven to be for so many readers, the obvious next step was to see whether he would be willing to perform comparable magic for the post–World War II environmental movement, half a century later. Happily, he was eager to take on the assignment, and the result is the book you hold in your hands. Like its predecessor, *The Environmental Moment* reprints short documents that quickly convey the myriad and sometimes unexpected roots from which modern environmentalism emerged: congressional legislation, presidential speeches, newspaper accounts of air pollution disasters, celebrations of wild nature, Stewart Brand's *Whole Earth Catalog*, Walt Kelly's famous Pogo cartoon, and so on and on. The pleasure of the collection lies in the rich insights and the teacherly skill—to say nothing of the good humor and playfulness— that enable Stradling to intuit just what a reader needs to understand the entangled relationships that helped create the environmental ways of thinking and political alignments we now take more or less for granted. Work your way through these pages, and you'll soon realize that beliefs and actions often labeled with the singular noun "environmentalism" are anything but unitary. Like all complex political and cultural movements, this one has been the result of many currents flowing together in what is better thought of as an entire watershed than a single river. I could not be more delighted that *The Environmental Moment* now joins David Stradling's earlier book in the Classic Texts series. Together, they provide a superb introduction to environmental politics in the twentieth century.

ACKNOWLEDGMENTS

This collection came together very slowly, over the course of many years and many, many conversations with scholars and archivists. I am particularly indebted to the excellent historians who helped me identify and locate documents related to topics and parts of the country with which I have little familiarity. Special thanks to Robert Gioielli, Michael Egan, Chad Montrie, Andy Kirk, Keith Woodhouse, Teresa Sabol Spezio, Mark Harvey, Tom Robertson, Finis Dunaway, Matt Klingle, Jeff Sanders, and James Hillegas—all of whom will recognize their contributions to this collection.

I have also relied on the good work of archivists who have helped locate and reproduce valuable documents. I am indebted to Ann Sindelar, reference supervisor at the Western Reserve Historical Society; Wendy Pflug of the University of Pittsburgh Archives; Dorothy Hazelrigg, curator at the South Carolina Political Collections at the University of South Carolina; Claudia R. Jensen, senior catalog librarian/archivist at the Denver Public Library; Jim Quigel, USWA archivist at the Pennsylvania State University Special Collections Library; William LeFebre, reference archivist at the Walter P. Reuther Library of Wayne State University; Kevin Fredette with the West Virginia and Regional History Collection of West Virginia University Libraries; Scott Daniels, reference librarian at the Oregon Historical Society; Harry Miller, reference archivist at the Wisconsin Historical Society; Jenny Mandel, archivist at the Ronald Reagan Library; Marilyn Scott, curatorial assistant at the Ohio State University Billy Ireland Cartoon Library & Museum; David Kordalski at the Cleveland *Plain Dealer*; and Vern Morrison at Cleveland State University.

Many people also helped me acquire and gain permission to use documents with which they have a personal relationship. These include Pete Kelly, president, Okefenokee Glee & Perloo, Inc.; Ed Zahniser and Ben Beach of the Wilderness Society; Dale Burk, who sent me a copy of his Bitterroot photograph; and Stewart Brand, who granted me permission to include the cover of the *Whole Earth Catalog*.

This collection exists because of the hard work, encouragement, and patience of Bill Cronon and Marianne Keddington-Lang. My deepest thanks to them and the important work they do at the University of Washington Press. Thanks too to Tim Zimmermann and Mary C. Ribesky at the University of Washington Press and to freelance editor Amy Smith Bell. Finally, thanks to Brittany Clair Cowgill, who typed many of the documents and helped me gather permissions when I had too little time to make much progress on my own.

I dedicate this book to Nina, my nature girl, who came along too late to join her sister, Sarah, and her mother, Jodie, on the dedication page of a previous book. The three of them sustain me.

THE ENVIRONMENTAL MOMENT, 1968–1972

INTRODUCTION

Over the course of the 1960s, a growing number of Americans became increasingly concerned about environmental issues. Old, intractable problems such as air and water pollution grew more acute—eyes stung in L.A.'s gray haze of smog; Lake Erie stunk of decomposing algal blooms and floating fish. A series of relatively new problems gained attention too, as radioactive fallout from atmospheric tests of nuclear bombs imperiled life around the globe; a host of toxic chemicals threatened human health and ecological diversity; and suburbs sprawled into farm fields and deep woods, compromising ecosystems and human communities. Strip mining in Appalachia and clear-cutting in western forests laid bare an eroding earth, and shocking images of the destruction suggested that all natural resources were finite and going fast. The public worried about the survival of countless culturally significant species, including the American alligator, bald eagle, and gray wolf, not to mention the prized species of Africa and Asia, such as tigers, leopards, and elephants. Late in the decade, after terrible years of drought and famine in India, Americans increasingly feared rapid population growth, not just in the United States, which indeed grew rapidly in the three decades after World War II, but also around the globe and especially in the world's poorest nations. The survival of humanity seemed at stake.

On top of all of this, a series of discrete events cast a shadow over what had seemed a bright and prosperous future just a decade earlier. In March of 1967 the oil tanker *Torrey Canyon* hit ground off Cornwall, breaking apart on the rocks and sending millions of gallons of oil into the Atlantic Ocean and eventually onto the beaches of England and France. Other environmental disasters quickly followed, including a major spill from an oil rig off the coast of Santa Barbara, which spoiled beaches and killed sea life. Just months later, in June of 1969, Cleveland's Cuyahoga River caught fire, a blaze that wasn't caught on film but eventually did capture the imagination of Americans who wondered how things could have gotten so bad. Just days after the Cuyahoga fire, the world began to realize the extent of an insecticide spill on the Rhine River, which caused a massive fish kill in Germany and the Netherlands. At that very moment, United Nations Secretary General U Thant warned that "the future of life on earth could be endangered."[1]

As the 1970s began, the United States was engulfed in an environmental crisis, during which much of the news involved the fate of the planet. Even the president of the United States, Richard Nixon, a conservative politician with a long history of public service and essentially no previous expression of environmental concern, began to embrace an environmental agenda. In his 1970 State of the Union Address, Nixon helped define the new decade: "The great question of the seventies is, shall we surrender to our surroundings, or shall we make our peace with nature and begin to make reparations for the damage we have done to our air, to our land, and to our water?" In laying out a comprehensive environmental agenda, Nixon, true to his conservative ideals, made clear the individual responsibilities all citizens shared: "Each of us must resolve that each day he will leave his home, his property, the public places of the city or town a little cleaner, a little better, a little more pleasant for himself and those around him."[2]

A month later Nixon delivered a more extensive speech to Congress outlining his environmental policies—a speech excerpted in this collection—but 1970 didn't become the pinnacle of the environmental moment because of political leadership. Instead, environmental activism came from all quarters; all parts of American society participated. The environmental moment was bipartisan. Environmental activists were young and old; some were radical and some conservative. Organized labor became involved, as did the upper management of the nation's largest companies. Some environmental protests were led by minorities and the working poor; other protests involved mostly middle-class whites and wealthy patricians. Some environmental activists worked to preserve wilderness, while others struggled to improve inner-city neighborhoods.

On April 22, 1970, the breadth of the growing environmental movement became clear with an outpouring of action and rhetoric on the first Earth Day, a nationwide teach-in on environmental issues. On college campuses, high school grounds, city streets and plazas, people of all kinds gathered to teach and learn about local, national, and global environmental problems. And many of them set to work that day, most notably by picking up trash in their communities. That first Earth Day revealed an intense national concern for the environment, an intensity not seen before or since. This collection of documents seeks to describe this environmental moment through the words of participants: activists, politicians, scientists, and average citizens. The diversity of voices heard throughout this volume captures the breadth and intensity of environmental concern in the years surrounding 1970.

THE DEVELOPING CRISIS

Environmental activism was not new in the late 1960s, of course. Modern environmentalism had deep antecedents, dating back at least to the late-nineteenth-century conservation efforts that had led to the creation of the first national parks, the establishment of the national forests, and the development of state park systems around the country. By the time 1960s activists took on a range of urban environmental problems, city dwellers had been working to improve urban environments for more than a hundred years—building sewer systems, improving garbage collection and street cleaning, as well as attempting to control coal smoke. The nation had already benefited from great environmental leaders, including Theodore Roosevelt, who among other things established the country's first wildlife refuges in the first decade of the twentieth century. In the 1930s conservation became a major national goal, with the Civilian Conservation Corps and other Depression-era programs planting billions of trees and developing recreational facilities on public lands around the country.

In the decades after World War II, however, environmental rhetoric took on a new urgency, perhaps as activists attempted to match the quickening pace of economic growth and technological change. The nation invested heavily in infrastructure in the postwar decades, most notably in building the interstate highway system. Consumers purchased cars and put them to great use in commuting and vacationing. Indeed, the increasing mobility of American tourists, many of whom drove deep into wild nature for hiking and camping, led to a growing concern about the fate of wilderness—defined principally as the absence of roads and cars—in the United States. At the same time, destruction in urban neighborhoods, as a result of highway building and urban renewal demolitions, led to increasing concerns for historic preservation. Altogether, the booming economy, industrial development, and the growing influence of automobile-reliant transportation led to a broad movement concerned with the preservation of wilderness, farmland, and urban communities.

Activism in the 1950s produced some positive results, but even more came in the 1960s. The antinuclear movement, led in part by scientists concerned about the troubling consequences of strontium 90 in the environment, met with real success in 1963, when the United States, the Soviet Union, and the United Kingdom signed the Limited Test Ban Treaty, prohibiting the explosion of nuclear devices in the atmosphere. The next year the long battle to preserve wild nature finally resulted in the creation of wilderness areas on federal lands. Soon thereafter, states began to use the Wilderness Act of 1964 as a model to create wilderness areas on state land too. In a narrower but telling struggle in the Hudson Valley, activists battled against Consolidated Edison's 1962 plan to build a power plant at Storm King, a scenic mountain that plunged into the river in the Hudson Highlands. The battle encouraged the development of more aggressive environmental activism and led to the creation of new interest groups, including Scenic Hudson and the Riverkeepers.

Despite these and other victories in the 1960s, the sense of crisis deepened as the decade progressed. The social context certainly contributed to the anxiety of the age, as concentrated urban poverty and racism helped fuel a series of horrific riots in inner-city neighborhoods around the country and antiwar protests turned violent. The brutality of the Vietnam War, seen nightly on the evening news, along with rising crime rates at home, fostered a sense of turmoil and confusion. Civilization itself seemed threatened. The urban and environmental crises came together in the minds of many Americans, including Allan Temko, the San Francisco–based architectural critic. In 1963, Temko described his anxiety in the *New York Times*: "Confronted by an environmental crisis of now almost incredible gravity as traditional urban civilization disintegrates without a coherent order of technological civilization to take its place, our supposedly affluent and inventive society finds itself strangely powerless to establish rational patterns of growth."[3] These sentiments spread as the decade wore on.

Temko's essay was just a small part of the national discussion of environmental problems. The late 1960s saw a dramatic increase in the publication of environmentally focused books, articles, pamphlets, government documents, and scientific literature. Many periodicals dedicated special issues to the environment or ecology—from the radical magazine *Ramparts*, which ran its "Ecology Special" in May 1970, to more mainstream outlets like *National Geographic*, which ran an issue on the topic of "Our Ecological Crisis" in December of that year. The rising attention to the environment in the press mirrored changes in the political sphere, where there was an outburst of environmental legislation at the state and federal levels, and in the scientific and engineering communities, where there were near-frantic searches for solutions to pressing and seemingly intractable problems.

This wasn't just an American phenomenon, of course. Environmental issues gained attention throughout the industrial word. *The Ecologist,* for instance, began publication in London in 1970, and its circulation reached around the globe. The popularity of this journal spoke to the growing appreciation of ecological science as a critical tool for understanding the problems facing humanity, and the rising awareness of the interconnectedness of natural ecosystems and human communities. At the same time, a greater number of scientists at research universities were undertaking important studies of natural communities and systems—many of which were experiencing significant disruption from human development. This was the age of ecology, when the industrial world increasingly turned to the "science of survival," as *Time* magazine called it in early 1970, to solve the world's most pressing problems.

Not surprisingly much of the published material on the environment described new scientific research or engaged more directly in the effort to influence public policy, but the environmental crisis had an even broader influence on American culture. Nature shows gained in popularity through the 1960s, including the Emmy-winning *Mutual of Omaha's Wild Kingdom,* which began in 1963, and *The Undersea World of Jacques Cousteau,* which began airing five years later. Significantly, both of these shows featured scientists at work in the natural world and both projected an environmentalist subtext. As the environmental moment peaked, radios too echoed the themes of the day. In September of 1970, Neil Young's popular song "After the Gold Rush" refrained: "Look at mother nature on the run in the 1970s." A year later, Marvin Gaye's "Mercy Mercy Me (The Ecology)" became a big hit, even though its lyrics addressed oil spills, radiation poisoning, and overpopulation—not exactly the staple of Motown Records. Even children's literature took an ecological turn with the publication of Dr. Seuss's fable of ecological destruction, *The Lorax,* in the fall of 1971. All of this together represented something new: a heightened cultural emphasis on the environment—and crisis.

Why did this happen in the years surrounding 1970? Several trends came together. First, both production and consumption rose rapidly in the postwar economy, and the growth came with too little forethought about wastes and harmful by-products. Regulatory structures simply could not keep pace in the age of abundance. Second, rapid technological change, including the development and use of nuclear bombs, atomic energy, and dangerous chemicals like DDT, all contributed to an increasingly hazardous world. In the United States the turn toward automobile travel was especially influential in sparking concern about the environment, as lead-laden exhaust hung over traffic-choked cities, and highways pushed through historical urban neighborhoods and into formerly serene rural communities. Rapid physical change to the American landscape clearly contributed to the sense of crisis in the late 1960s.

The political culture of the 1960s also contributed to the creation of the environmental moment. As baby boomers came of age, they questioned authority, both economic and political. American politics became increasingly participatory, and citizens did much more than vote periodically to express their political preferences. This was a decade of marches, protests, and teach-ins. Direct action became a favored means of democratic participation. Environmental activists had clearly learned from the techniques of the civil rights movement and Vietnam War protesters. Politicians learned too, and government became increasingly responsive to popular demonstrations.

The sense of crisis was also fueled by the rhetoric coming from environmental interest groups. The 1960s was a decade of intense interest-group politics, as such established groups as the National Audubon Society and the Sierra Club as well as many new ones—like the Environmental Defense Fund (1967), Friends of the Earth (1969), and Greenpeace (1971)—made certain that environmental issues remained on government agendas. Some of these groups were international from the outset, including Greenpeace, and they had truly global agendas, but concerned citizens also created local organizations to solve specific problems. This was the case for Group Against Smog and Pollution (GASP), founded in Pittsburgh in 1969, and the Citizens Association of Beaufort County—both of which produced documents included in this collection.

Many Americans think of themselves and their nation as having a special connection to the natural world, especially the wilderness of the American West. This fixation on wild nature is part the heritage of a frontier nation, but the movement to preserve wilderness cannot explain the development of the environmental moment. Wilderness protection certainly informed American environmentalism, but the late 1960s surge in environmental activism spanned the industrial world. Citizens in Western Europe and Japan, in particular, demanded improved environmental quality. Each country has its own political culture, which influenced the shape and results of environmentalism in each nation, but there was a remarkable concurrence in timing and effectiveness of governmental responses. For example, 1970 witnessed the creation of the Environmental Protection Agency in the United States as well as the creation of Japan's Central Pollution Countermeasures Headquarters. The next year, France created its Ministry of Environment. In addition, the

Council of Europe declared 1970 the European Conservation Year. On three continents governments became more interventionist, revealing the strong connection between industrial development and the need for environmental regulation.

Although this volume concerns the American environmental movement, it also contains evidence of the international nature of this moment. In 1968 the United Nations began preparations for a global environmental conference, and four years later the United Nations Conference on the Human Environment took place in Stockholm. In the intervening years, governments around the world, environmental interest groups, and scientists in a wide variety of fields prepared study after study on environmental issues. It is no coincidence that the years in which the world prepared for and participated in its first global environmental conference are the years that mark the environmental moment in the United States.

QUESTIONS FOR THE DOCUMENTS

The documents in this collection can help us answer two different types of questions. First, there are the questions that faced Americans in the environmental moment: What caused the environmental crisis? Should we blame broad forces, like capitalism, industrialization, or perhaps Christianity? Were specific corporations or industries to blame, or maybe just average consumers? Perhaps these documents might help answer more fundamental questions: How serious were these problems? Was this truly a crisis, or were some activists just alarmists? Some problems clearly existed, of course, but even in instances where some agreement might exist as to the problem, debates existed on how to find solutions. And so, we might ask, as many did at the time, what resources should be brought to bear to solve environmental problems and, just as important, what sacrifices should be made, and by whom?

These documents might also help us answer the questions that historians now ask about the environmental movement. Most basically, how should we define "environmentalism"? Whom should we categorize as "environmentalists"? Not everyone who expressed concern for the environment, or even became active in efforts to improve or protect the environment, accepted the label "environmentalist." This was true of Stewart Brand, the appropriate technology advocate who started *The Whole Earth Catalog* in 1968—the cover of the second edition appears among the documents—and inspired a movement for responsible technology designed to decrease environmental destruction. Like many activists in the late 1960s, Brand thought "environmentalist" was much too narrow a term to describe his own interests. He, like others, had long associated the term with wilderness advocates. But we need not be so narrow with our use of the term.

Surely many environmental activists fit the stereotype of middle-class whites concerned largely with recreational opportunities, such as backpacking, backwoods camping, and whitewater rafting. Many environmental activists were members of organizations dedicated to the protection of wild places, such as the Sierra Club and the Wilderness Society. The documents included in this collection, however, reveal a real diversity among those who spoke out about environmental threats. Participants in the environmental moment were not just the stereotypical young, liberal Californians or, put more cynically, the Berkeley radicals in search of a cause or in search of protection for their privileged lives. In reading these documents, we might ask why subsequent portraits of environmentalists have cast them so narrowly. Given the range of the authors and perspectives represented in this collection, we must ask if "environmentalist" should be a more inclusive term. These documents reveal that environmental skeptics were not always those who had a direct economic stake in the status quo or greedy corporate executives and their political allies. We might also ask why subsequent portraits of skeptics have cast them so narrowly as well.

Finally, although these documents do not address the question directly, perhaps there are clues here as to why the sense of crisis waned, why the environmental moment was relatively brief. Surely environmental activism has persisted and environmental regulation has intensified, but the environmental moment did pass, despite periodic rebirths in the movement. Through the 1970s the sense of urgency diminished, as did the level of national engagement with environmental issues. Economic troubles, especially related to deindustrialization and inflation, diverted attention from environmental issues. In addition, the late 1970s and 1980s witnessed an increasingly troubled relationship between science and society, with average Americans growing suspicious of scientific data and the conclusions of the scientific community. Furthermore, increasing skepticism about politics and politicians has helped create an atmosphere of decreased confidence in government as a problem solver. Did the rhetoric of crisis itself contribute to the rapid diminishment in the breadth of public concern? Did it play a role in the rise of the New Right and the effective critique of government bureaucracy?

THE DOCUMENTS

I have edited these documents for continuity, with limited elisions, so that readers can take the documents on their own terms, rather than mine. Along the way I corrected some typographical errors and perhaps introduced others. I have grouped the documents in five parts. The first, "Warnings," gives just a small sampling of the environmental literature that preceded the most active years of the environmental movement. I intend the similarly brief last part, "Continuation," to serve as a reminder that environmental problems continued to surface and environmental activism remained alive, even after the environmental moment began to fade.

The majority of the documents are in the middle three parts of the book, where they appear in rough chronological order. This separation may introduce some artificial distinction among them, because they really represent a continuity of concern over these five years, but the division allows readers to pause and reflect on a few documents at a time. I have not clustered the documents around specific issues or particular places. Gathering documents under such headings as "wilderness" and "air pollution" would impose an artificial order, suggesting that environmental issues appeared on the scene one at a time and were taken on as such, held apart in mind and action by activists and politicians.

The documents in this collection can hold different meanings for different readers, and hence they can suggest a wide variety of meanings for the environmental crisis. Taken together, however, they teach one unmistakable lesson: that all of these issues came rushing together—in the press, in the public consciousness, and onto political agendas. That is why we can rightfully call these few years the environmental moment.

NOTES

1 Quoted in "The Deteriorating Environment," *New York Times*, June 25, 1969.
2 Richard Nixon, State of the Union Address, delivered on January 22, 1970, *The American Presidency Project*, online at http://www.presidency.ucsb.edu/ws/index.php?pid=2921&st=&st1=#axzz1KZFcZWFq.
3 Allan Temko, "Things Are Getting Too Crowded, Too Mechanized, and Too Noisy," *New York Times*, October 13, 1963.

PART 1

WARNINGS

IN THE DECADES AFTER WORLD WAR II, A VARIETY OF AUTHORS—
scientists, politicians, and activists—produced an increasingly urgent literature on the environmental consequences of economic and demographic
growth. Many writers began to question the very idea of "progress," and especially the widely held belief that the American system of capitalism would
produce ever-increasing wealth. In the face of growing evidence concerning
radioactive fallout from nuclear testing and the potentially ecologically crip-
pling effects of commonly used toxic chemicals, skeptical voices questioned
the nation's faith in science and technology.

This part of the book includes documents that warn of impending environmental disaster. Each one urges action. The first document was not widely
read, but it concerns an important event—the smog disaster at Donora, Pennsylvania, in the fall of 1948. Even though few people encountered the federal
government's Public Health Service bulletin concerning the disaster, Donora
attracted lasting public attention, and Americans began to recognize that air
pollution was more than a nuisance. South of Pittsburgh on the Mononga-
hela, Donora had a population approaching fourteen thousand, many of them
Slavic immigrants employed in the steel and zinc plants along the river. During the deadly smog event, steep hillsides and a temperature inversion trapped
pollution for five days at the end of October, and more than fourteen hundred residents reported severe effects, mostly shortness of breath and stinging
eyes. Later research, conducted by the Division of Industrial Hygiene, concluded that perhaps twenty deaths could be attributed to the smog. The report
excerpted in this book describes the "Air Pollution in Donora, PA" in part by

describing the cases of all of the hospitalized victims of the smog. Even though these descriptions, just a few of which are included here, spoke to the severity of the problem, the recommendations made at the end of the document speak to a very limited concept of the growing air pollution problem in the United States.

Ten years after the Donora disaster, Paul Shepard wrote an influential essay that appeared in *The Atlantic Naturalist*, a publication of the Audubon Society of Washington, D.C. At the time Shepard was just beginning a long career of studying the human relationship to the natural world; he went on to become a leading thinker in the Deep Ecology movement. In "The Place of Nature in Man's World," Shepard used some older language about "conservation problems," but the problems he describes—related to such toxic chemicals as DDT and strontium 90—were quite new. This 1958 essay reveals a young man making the intellectual journey that many Americans would take as they thought more ecologically about the human relationship to nature. Shepard feared a world awash in poisons, a world that is "not quite fatal," where urban residents wear masks to protect themselves from toxic air.

The Wilderness Society's Howard Zahniser expressed a different set of concerns as he addressed the seventh biennial wilderness conference in San Francisco in 1961. Like Shepard, Zahniser was thinking of the long term, but his life-long concern was wilderness, not gas masks for city dwellers. Also like Shepard, Zahniser was an important environmental thinker and author. Indeed, just three years after this speech in San Francisco, Zahniser helped push through Congress the Wilderness Act, of which he was the primary author. In this oft-reprinted speech, Zahniser used some of the language that appeared in the act, including the definition of wilderness as areas where "the earth and its community of life are untrammeled by man." In arguing for wilderness, Zahniser also argued for a new definition of progress.

The fourth document included here is by far the most important. The scientist and gifted writer Rachel Carson published her seminal work, *Silent Spring*, in 1962. Almost instantly it became essential reading for anyone concerned about the environment, and it has entered the cannon of great American books. Carson's immediate concern focused on the reckless use of very toxic pesticides, most infamously DDT, but *Silent Spring* also raised broader philosophical questions about the human inclination to dominate nature, to attempt to control the natural world through violence or poison. The first chapter, included here, offered a frightening "Fable for Tomorrow," envisioning a world greatly diminished by "white granular powder" dropped from above. Carson's writing was so effective in part because she knew how to ask the right questions. In her second chapter, "The Obligation to Endure," she asks why we would take such risks in spraying deadly chemicals so liberally.

The last document in this part of the book concerns another growing threat: industrial and residential growth. In the early 1960s Consolidated Edison proposed building a power plant at the scenic Storm King Mountain on the shores of the Hudson River. A number of conservation organizations protested vehemently and some of them gathered to create a new group, which became known as Scenic Hudson. Among these activists was the author Carl Carmer, who had written the Hudson volume of the popular Rivers of America series. In 1964, Carmer testified at a hearing before the Federal Power Commission, which by law had to review and approve Con Ed's plan. Carmer described a broader fear felt along the Hudson, not just of an obtrusive power plant but of a "wild irresistible monster" (development) that would suck the beautiful and historical valley into "the maw of the city." Like Zahniser and Carson, Carmer asked his audience to reconsider the meaning of progress.

AIR POLLUTION IN DONORA, PA

EPIDEMIOLOGY OF THE UNUSUAL SMOG EPISODE OF OCTOBER 1948, PRELIMINARY REPORT

FOREWORD

This study is the opening move in what may develop into a major field of operation in improving the Nation's health. We have realized, during our growing impatience with the annoyance of smoke, that pollution from gases, fumes, and microscopic particles was also a factor to be reckoned with. But it was not until the tragic impact of Donora that the Nation as a whole became aware that there might be a serious danger to health from air contaminants.

Before the Donora episode, there had been only one other similar incident in history. In 1930, in the Meuse Valley of Belgium, a period of intense fog in a heavy industrial area resulted in the death of 60 persons. Although several studies were made of those fatalities, the Donora study is the first thorough investigation into every facet of an air-pollution problem, including health effects as well as deaths.

The Donora report has completely confirmed two beliefs we held at the outset of the investigation. It has shown with great clarity how little fundamental knowledge exists regarding the possible effects of atmospheric pollution on health. Secondly, Donora has emphasized how long-range and complex is this job of overcoming the problem of air pollution—after we get the basic knowledge of its effects. This intensive piece of work by the Division of Industrial Hygiene of the Public Health Service will have its greatest value as the blueprint for our plan of proceeding to get that knowledge.

Our first step now, of course, is immediate basic research. We need to investigate for instance, what long-range effect continued low concentrations of polluted air has on the health of individuals—not only healthy individuals,

Public Health Bulletin No. 306, 1949, excerpts from pages iii, 31, 32, 36, 37, and 165.

but those with chronic diseases and the aged and children. We know nothing about the indirect effect of air pollution on persons with diseases other than those of the respiratory tract. We also need immediate research into another indisputable effect of air pollution: its ability to shut out some of the healthful rays of the sun.

When we find the answers to all of these unknowns, we can proceed to the problem of eliminating the causes. As a proof that air pollution is a health matter, as a model for future studies in air pollution, and as an important phase of our increasing efforts in the field of environmental health, this study will be invaluable.

Leonard A. Scheele
Surgeon General

HOSPITALIZED PERSONS

About 50 persons were hospitalized during the smog period; the records of 32 were obtained to study certain phases of the acute smog illness not available to us from other sources. The hospital records were in various stages of completeness from the point of view of studies made of the patients. The cases are presented in detail at the end of this section. The group consisted of 25 males and 7 females with ages ranging from 8 to 76 years. More than two-thirds of these persons were over 55 years old. There were three non-white persons hospitalized. . . .

Case A-1, age 56, white, male, married, born in Czechoslovakia. In 1909 he came to Donora where he worked in the wire plant until his retirement 8 years ago because of heart disease.

His acute illness began in the morning of S-day [Wednesday, October 27], with dyspnoea [shortness of breath], orthopnoea [the inability to breath while lying flat], and a productive cough. Since there was no improvement, he was hospitalized on day No. 1.

The physical examination revealed that he was in acute respiratory distress. His head and neck were negative except for cyanosis of the mucous membranes [a blue hue caused by lack of oxygen]. . . . He was placed in an oxygen tent upon admission to the hospital and was given 2 ml. of aminophyllin intramuscularly, sedatives, and a liquid diet. Within 2 days (on day No. 3) his dyspnoea improved, cough diminished, and cyanosis disappeared. . . .

Case A-2, age 47, white, female, married, housewife, born in Czechoslovakia. She came to the United States in 1921 and lived in Donora since 1926.

She became acutely ill on the morning of S-day with a sense of painful constriction in her chest. The chest pain became worse that evening and she was unable to sleep. On day No. 1 she became dyspnoeic, orthopnoeic, cyanotic, and developed a nonproductive cough. Despite medical attention her symptoms persisted and she was hospitalized on day No. 2. . . .

She vomited clear fluid several times during the first hospital day. She then made a rapid recovery. . . .

In an interview at a later date the patient stated that she also had the following complaints during the acute illness: She detected a foul odor; had an acid taste, headache, and weakness. The weakness persisted for 2 weeks. . . .

Case A-23, age 47, Negro, male, married, steel plant worker, born in the United States, was hospitalized on day No. 4 after becoming ill on day No. 2.

On physical examination he was found acutely ill and semi-comatose. The examination of the chest revealed a few scattered rales anteriorly and posteriorly, and limited expansion of the chest. . . .

He was given penicillin and placed in an oxygen tent. He made a rapid recovery and was discharged on the third hospital day with a diagnosis of bronchial asthma and tracheobronchitis.

Case A-29, age 65, white, male, widower, born in Czechoslovakia, resident of Webster, coal miner and zinc plant worker, was hospitalized on day No. 14.

His illness began on the morning of day No. 3 with dyspnoea, chest pain, and ankle oedema, which progressed after that date. He was a known bronchial asthmatic. . . .

He was placed in an oxygen tent and given penicillin. He did not respond to the therapy but became progressively worse. He died at 10:45 p.m. on December 22, 1948.

RECOMMENDATIONS

1. Reduce the gaseous contaminants especially sulfur dioxide and particulate matter discharged from the sinter plant Cottrell stacks.
2. Reduce the particulate matter and carbon monoxide from the zinc smelters.

3. Reduce the particulate matter and sulfur dioxide discharged from the waste heat boiler stacks.
4. Reduce the discharge of oxides of nitrogen and acid mists from Gay-Lussac stacks.
5. Reduce the amount of particulate matter and carbon monoxide from the waste blast furnace gas.
6. Reduce the amount of carbon monoxide discharged from the stove and sinter stacks.
7. Reduce the amount of particulate matter discharged from the sinter plant and open hearth stacks.
8. Reduce the amount of particulate matter discharged from the waste heat and blast furnace boilers and the sulfur dioxide from the waste heat, steel and wire plant boilers.
9. Reduce the amount of particulate matter discharged from domestic heating systems, steam locomotives and steamboats.
10. Establish a program of weather forecasts to alert the community of impending adverse weather conditions so that adequate measures can be taken to protect the populace.

PAUL SHEPARD

THE PLACE OF NATURE IN MAN'S WORLD

THE ATLANTIC NATURALIST (APRIL 1958)

The place of nature in man's world is a title borrowed from E. G. Murray of McGill University. He was describing to the St. James Literary Society the enormous impact of the smallest microbes on the destiny of man. There is a hint of sarcasm in such a title, in the image of nature shifting for herself in the humanized landscape. It shows the naturalist's rueful recognition of topsy-turvy values in a technological society. Its irony stems from the deep conviction that we do not change natural laws nor conquer nature, only impose ourselves on it and, in the end, suffer for whatever damage we do.

Of the two great conservation problems now before us—a population explosion and the poisoning of our environment—the second is a question of our direct action in nature. It is amenable only to an approach which does not deal piecemeal with resources. The biochemistry of this corruption is much too fundamental to isolate itself in soil, water, or wildlife. Pollution is just as much a product of capsule thinking in conservation as it is of exploitive disregard for the future and the habitat.

Our environment continues to become a more poisoned place. The self-cleansing capacity of the air, water, and soil is all that has saved us from the fate of a yeast population in a vat of cider—where it manufactures alcohol until it poisons itself. Mere diffusion and dilution would long ago have ceased to absorb our production of toxic wastes were there not compensatory, purifying natural machinery. According to the Department of Agriculture, we are still adding silt to the air and to streams in volumes comparable to the 1930's. Erosion is not only a problem of topsoil loss, it is a pollution problem. There is serious doubt expressed by the U.S. Department of Public Health that the

Excerpts from pages 85–88. Used with permission from the Audubon Naturalist Society, available at http://www.audubonnaturalist.org

self-cleansing ability of the atmosphere will stand up under the new addition of industrial wastes. Our larger cities now wear a permanent gray cap of toxic gasses and materials in suspension. These substances are chemically active and subject to transformation by radiation, as vividly indicated in the annals of recent outbreaks of respiratory ailments. Aside from the enormous bulk of combustion residue, there are organic materials from the plastics and other industries, and volatile acids which kill vegetation as well as animals. The country is 6,600 municipal sewage units short, and lacks 3,500 industrial units. Oil on the sea kills thousands of sea birds and other life.

Added to the soil and water, as well as the air, we now have continuing use of the traditional metallic poisons, plus the new organic insecticides, the chlorinated hydrocarbons, such as DDT, and the organic phosphates, such as parathion. The latter are so powerful that in pure form the minimum dose to kill a rat cannot be measured. DDT, which promised ten years ago to rid every dairy barn of horseflies, is no longer being used around dairies partly because the flies have become immune, but mostly because it accumulates and its toxicity is undiminished after passage through the cow in whose milk it appears. It is now annually sprayed over cities from New Haven to St. Louis to kill bark beetles and has been blanketed over extensive forest areas in the West and North.

The pollution of the atmosphere as high as 40,000 feet has taken a new twist with the radioactive residues of nuclear explosion, debris which comes down with gravity, with rain, and with dust. Unlike the other poisons, this new one yields inheritable effects. But like them, its danger is regarded in terms of its intensity of application. The concept of a threshold dosage, the maximum permissible amount that a human can take without pathological symptoms, is the crux of a score of running controversies, from the consumption of DDT in meat and vegetables to the absorption of a strontium-90 from fallout.

The defect in the expiatory thinking behind a "maximum safe dosage" is, first, that the more we know the more it is revised downward. A "safe" dose of DDT yesterday may poison our livers tomorrow. "Safe" roentgens received by X-ray technicians earlier are now known to have increased their chances of defective offspring. Secondly, it is wrong because it idealizes life with only its head out of water, inches above the limits of toleration of the corruption of its own environment. Why should we tolerate a diet of weak poisons, a home in insipid surroundings, a circle of acquaintances who are not quite our enemies, the noise of motors with just enough relief to prevent insanity? Who would want to live in a world which is just not quite fatal?

Under such conditions we might welcome the final modicum which pushes us over the brink.

The third and perhaps most important error in a "maximum permissible dosage" is evident in the question, "dosage for whom?" What is the maximum safe dose for the bacteria which live in the soil, whose numbers and manifold activities there are essential to the life of the soil? What is the maximum safe dose for the insect- and therefore insecticide-catching swallow, which is not essential to the soil, to the air, or to human activity? To base the concentration of poisons on the levels which man and his domestic plants and animals can tolerate could mean, in the coming world of chemical technology, the climax of the drama of separating the "useful" from the "useless." Even Noah did not presume to exclude any species from the ark.

The middle ground between that fringe element who would insist that we share our tomatoes with every worm in the garden and those who engineer the natural world out of existence requires a new conservation approach. It admits acceptance and enthusiasm for the triumphs of human ingenuity, taken with the understanding that the capacity to get into trouble is thereby increased. It requires a new unit of thinking about the subject matter with which conservation deals. Production techniques—agronomy, silviculture, mainstream engineering, predator and pest control—none in the past has been characterized by what Albert Schweitzer has called a "reverence for life." The most severe critic of the notion that these activities are conservation is Joseph Krutch, who says, "What is commonly called 'conservation' will not work in the long run, because it is not really conservation at all but rather, disguised by its elaborate scheming, a more knowledgeable variation of the old idea of a world for man's use only." Yesterday toadstools had no value, except for their weird beauty, appreciated only by such eccentrics as Henri Fabre, the naturalist. We would just as soon have crowded them out of the fields and forests. Then it was discovered that the fungi are necessary for the root growth of trees; that they produce antibiotics; that their saprophytic action is important in the metabolism of the soil. Many young men and women being trained as engineers, forester, farmers, or wildlife managers would be embarrassed to admit to their colleagues or faculty that they enjoy the beauty of the "useless" creatures around them, or that these are of any real importance in the execution of their profession. Many of them adopt the specious and convenient fiat of dividing nature into blacks and whites.

One sympathizes with the chemical company official who said in a speech not long ago that he would like to withhold every chemical from the market until every possible effect upon all organisms was known, but that mechanical brains had calculated the amount of research necessary and found it prohibitive. Does exoneration come by confession of good will, by energetic experimental testing, and by the presence of the word CAUTION in large print on the label? Any reasonable person knows that we will continue to use chemicals, and the liability for the makers and the users does not end in the test plot, or with its effect only on crops. Although we cannot know all the possible effects, we do recognize the symptoms of a natural community that is sick or a watershed in difficulty. The president of another large chemical company, the Thompson Chemicals Corporation, had this to say, "We have decided to withdraw entirely from the production, distribution, and research of the *presently known* agricultural insecticides.

"A twelve-year study has convinced us that the currently known and used broad spectrum insecticides and their wide-scale application to agricultural crops—although giving temporary control and temporarily increased yields— are at best palliative, and perhaps will prove dangerous and un-economic in the long run.

" . . . The growing number of insect pests of economic importance that are becoming resistant to presently used agricultural insecticides demonstrates a serious inherent danger in the side-scale use. The imbalance of the fauna population caused by the destruction of the natural predators and parasites (thus allowing the uninhibited development of the insect pests) is further proof to us of the unsoundness of the current chemical insecticides. This cannot only result in rapid and dynamic developments of the insect pest from the few not controlled by the application, but can easily cause heretofore unimportant insects to increase to the status of economic pests, once the predator-parasite balance has been upset.

" . . . The ingestion of presently employed insecticide residues by humans and other warm-blooded animals is a correlative problem of a *highly serious* nature. The industrial hazards inherent in the indiscriminate, wide-scale application of chemicals of such highly toxic nature also causes concern."

The uninhibited movement of poisons in currents of air and water, and their accumulation over periods of time, make the health of the landscape itself and the survival of whole populations the measure of success. Those long-term and widespread effects are problems which must be faced by manufacturers of poisons and the land-management professions in which they are used. The same is true of public officials in the disposal of city sewage and fumes: the water does not end at the city limits.

It seems reasonable that a middle ground between the fanatic wearing of masks to keep from killing gnats and the imperious destruction of nature requires the prevention of extinction of any species, the maintenance of areas with natural vegetational types as cleansing filters of air and water, and the safeguarding of natural processes which tend to dampen the fluctuations of natural populations and to flush the accumulated poisons. . . .

HOWARD ZAHNISER

WILDERNESS FOREVER

SPEECH DELIVERED AT THE SEVENTH BIENNIAL
WILDERNESS CONFERENCE, SAN FRANCISCO, 1961

It is a bold thing for a human being who lives on the earth but a few score years
at the most to presume upon the Eternal and covet perpetuity for any of his
undertakings.

Yet we who concern ourselves with wilderness preservation are compelled
to assume this boldness and with the courage of this peculiar undertaking of
ours so to order our enterprise as to direct our efforts toward the perpetual—
to project into the eternity of the future some of that precious unspoiled ecological inheritance that has come to us out of the eternity of the past.

This is a requisite of our undertaking, and there is yet another of primary
importance also:

We must deal with actual areas. Only as we preserve areas of wilderness
does there exist in reality the basis for a vital interest in all the many aspects of
wilderness that give it the meanings we have been discussing, not only in our
recreation but also in our science, literature, art, entertainment—our whole
culture, our way of living.

We who are concerned with wilderness preservation must accordingly
have these two clear purposes: We must relate all our effective concerns and
efforts to the preservation of actual areas, and we must work for their preservation in perpetuity.

When we address ourselves to wilderness preservation with such a purpose, we are dealing with those still remaining areas of the earth where the
landscape is not dominated by man and his works, areas where the earth and

As found in David Brower, ed., *Wilderness: America's Living Heritage* (San Francisco:
Sierra Club, 1961), excerpts from pages 155–56 and 160–62. Used with the permission of
the Wilderness Society.

its community of life are untrammeled by man, where man himself is a member of the natural community, a wanderer who visits but does not remain, whose travels leave only trails.

These are the areas that still retain their primeval environment and influence, that remain free from routes that can be used for mechanized transportation, where the freedom of the wilderness still lives on unfettered by the restrictions of the urban industrial life to which mankind has become increasingly confined, primeval areas where a human being can still face natural conditions directly without the mediating conveniences and instruments of domination fashioned in his inventive and technological civilization.

These are the areas that are still as God has been making them without man's aid, but for the protection of which the Almighty now seems to be relying on this His remarkable creature, man—this free-willed, so often intractable participant in the eternal purposes of the whole boundless universe. . . .

We must go out from our conferences on wilderness to work with our people on the wilderness—to inform them through the press, television, through all our media of publicity, and to help them organize in such a way as to make their informed purposes effective.

The question is not one of dealing or not dealing with all our various and far-flung problems. It is a question of how we shall deal with them in our development of an enduring program.

It is of great importance to enlist the civic leaders of our communities in the study of problems relating to wilderness preservation. In every community there are among the local people, businessmen, teachers, clergymen, laborers, farmers, and the many other groups, those who will become effectively interested in wilderness if we can only help them get started.

This is a leadership task that involves us in showing people how to provide a positive influence under the prerogatives of our democratic system.

In brief we need to practice the art of helping others work effectively in fighting for the things in which we believe.

Nor have we exhausted our educational possibilities when we have adapted our wilderness information activities to all the common media of information and means for organizational effectiveness. We need also to entertain in our own imagination every new stimulus of the public mind that we perceive.

The pioneer spirit that stirs in youth is the spirit of the wilderness. Through wilderness experience it can be reborn. We can stir again the youthful energy which has made America strong. We can show that there are yet new frontiers, including our own frontier in fashioning a wilderness program that will endure.

Primeval wilderness, once gone, is gone forever; but it can be preserved forever. The vision of generation after generation, through an enduring future perpetuating a soundly established human purpose, is as glorious as a man's view of sons and daughters when he himself senses the period of his own time and cherishes more and more the Eternal.

The practical program for wilderness preservation, even in its discussion, leads us thus into the inspiring contemplation of something that endures. That is the nature of wilderness and we can hardly fail to realize it. What we must also recognize is that there is still the drive of the self-interest that exploits the wilderness for profit. There still are mining and lumbering interests who seek to confound, frustrate, and defeat every effort to secure wilderness as wilderness. There still are hazards in various enterprises that would continually modify wilderness rather than limit or regulate their own projects. We must use our inspirations to deal patiently, persistently, but practically with these contending forces.

Our political realities are such that we must continue, in our role as citizens, to strive to see the nation of which we are citizens espouse this cause to which we have become devoted. In this effort we are compelled to recognize that we must have the concurrence of many who have not yet or have not long shared our purposes. We must recognize that wilderness as a resource of the people has not been assured perpetuity until those among the people who would and could destroy it have been enlisted in or reconciled to its preservation. We must continue to work for the passage of the basic legislation that is the first step in whatever we can accomplish, and as it is enacted we must promptly mobilize for the ten- or fifteen-year program that it will inaugurate. There must not be any hesitancy in this, our immediate course of action.

If some of us may indeed become wearied physically, and profoundly, in the years through which frustrations continue—

Who are only undefeated
Because we have gone on trying—

we should never lose heart. We are engaged in an effort that may well be expected to continue until its right consummation, by our successors if need be. Working to preserve in perpetuity is a great inspiration. We are not fighting a rear-guard action, we are facing a frontier. We are not slowing down a force that inevitably will destroy all the wilderness there is. We are generating another force, never to be wholly spent, that, renewed generation after generation, will be always effective in preserving wilderness. We are not fighting progress. We are making it.

We are not dealing with a vanishing wilderness. We are working for a wilderness forever.

RACHEL CARSON

SILENT SPRING

1. A FABLE FOR TOMORROW

There was once a town in the heart of America where all life seemed to live in harmony with its surroundings. The town lay in the midst of a checkerboard of prosperous farms, with fields of grain and hillsides of orchards where, in spring, white clouds of bloom drifted above the green fields. In autumn, oak and maple and birch set up a blaze of color that flamed and flickered across a backdrop of pines. Then foxes barked in the hills and deer silently crossed the fields, half hidden in the mists of the fall mornings.

Along the roads, laurel, viburnum and alder, great ferns and wildflowers delighted the traveler's eye through much of the year. Even in winter the roadsides were places of beauty, where countless birds came to feed on the berries and on the seed heads of the dried weeds rising above the snow. The countryside was, in fact, famous for the abundance and variety of its bird life, and when the flood of migrants was pouring through in spring and fall people traveled from great distances to observe them. Others came to fish the streams, which flowed clear and cold out of the hills and contained shady pools where trout lay. So it had been from the days many years ago when the first settlers raised their houses, sank their wells, and built their barns.

Then a strange blight crept over the area and everything began to change. Some evil spell had settled on the community: mysterious maladies swept the flocks of chickens; the cattle and sheep sickened and died. Everywhere was a shadow of death. The farmers spoke of much illness among their families. In the town the doctors had become more and more puzzled by new kinds of sickness appearing among their patients. There had been several sudden and unexplained deaths, not only among adults but even among children, who would be stricken suddenly while at play and die within a few hours.

There was a strange stillness. The birds, for example—where had they gone? Many people spoke of them, puzzled and disturbed. The feeding stations in the backyards were deserted. The few birds seen anywhere were moribund; they trembled violently and could not fly. It was spring without voices. On the mornings that had once throbbed with the dawn chorus of robins, catbirds, doves, jays, wrens, and scores of other bird voices there was now no sound; only silence lay over the fields and woods and marsh.

On the farms the hens brooded, but no chicks hatched. The farmers complained that they were unable to raise any pigs—the litters were small and the young survived only a few days. The apple trees were coming into bloom but no bees droned among the blossoms, so there was no pollination and there would be no fruit.

The roadsides, once so attractive, were now lined with browned and withered vegetation as though swept by fire. These, too, were silent, deserted by all living things. Even the streams were now lifeless. Anglers no longer visited them, for all the fish had died.

In the gutters under the eaves and between the shingles of the roofs, a white granular powder still showed a few patches; some weeks before it had fallen like snow upon the roofs and the lawns, the fields and streams.

No witchcraft, no enemy action had silenced the birth of new life in this stricken world. The people had done it themselves.

This town does not actually exist, but it might easily have a thousand counterparts in America or elsewhere in the world. I know of no community that has experienced all the misfortunes I describe. Yet every one of these disasters has actually happened somewhere, and many real communities have already suffered a substantial number of them. A grim specter has crept upon us almost unnoticed, and this imagined tragedy may easily become a stark reality we all shall know.

What has already silenced the voices of spring in countless towns in America? This book is an attempt to explain.

2. THE OBLIGATION TO ENDURE

The history of life on earth has been a history of interaction between living things and their surroundings. To a large extent, the physical form and the habits of the earth's vegetation and its animal life have been molded by the environment. Considering the whole span of earthly time, the opposite effect, in which life actually modifies its surroundings, has been relatively slight. Only within the moment of time represented by the present century has one species—man—acquired significant power to alter the nature of his world.

During the past quarter century this power has not only increased to one of disturbing magnitude but it has changed in character. The most alarming of all man's assaults upon the environment is the contamination of air, earth, rivers, and sea with dangerous and even lethal materials. This pollution is for the most part irrecoverable; the chain of evil it initiates not only in the world that must support life but in living tissues is for the most part irreversible. In this now universal contamination of the environment, chemicals are the sinister and little-recognized partners of radiation in changing the very nature of the world—the very nature of its life. . . .

To adjust to these chemicals would require time on the scale that is nature's; it would require not merely the years of a man's life but the life of generations. And even this, were it by some miracle possible, would be futile, for the new chemicals come from our laboratories in an endless stream; almost five hundred annually find their way into actual use in the United States alone. The figure is staggering and its implications are not easily grasped—500 new chemicals to which the bodies of men and animals are required somehow to adapt each year, chemicals totally outside the limits of biologic experience. Among them are many that are used in man's war against nature. Since the mid-1940's over 200 basic chemicals have been created for use in killing insects, weeds, rodents, and other organisms described in the modern vernacular as "pests"; and they are sold under several thousand different brand names. . . .

The whole process of spraying seems caught up in an endless spiral. Since DDT was released for civilian use, a process of escalation has been going on in which ever more toxic materials must be found. This has happened because insects, in a triumphant vindication of Darwin's principle of the survival of the fittest, have evolved super races immune to the particular insecticide used, hence a deadlier one has always to be developed—and then a deadlier one than that. It has happened also because, for reasons to be described later, destructive insects often undergo a "flareback," or resurgence, after spraying, in numbers greater than before. Thus the chemical war is never won, and all life is caught in its violent crossfire.

Along with the possibility of the extinction of mankind by nuclear war, the central problem of our age has therefore become the contamination of man's total environment with such substances of incredible potential for harm—substances that accumulate in the tissues of plants and animals and even penetrate the germ cells to shatter or alter the very material of heredity upon which the shape of the future depends....

All this has been risked—for what? Future historians may well be amazed by our distorted sense of proportion. How could intelligent beings seek to control a few unwanted species by a method that contaminated the entire environment and brought the threat of disease and death even to their own kind?...

It is not my contention that chemical insecticides must never be used. I do contend that we have put poisonous and biologically potent chemicals indiscriminately into the hands of persons largely or wholly ignorant of their potentials for harm. We have subjected enormous numbers of people to contact with these poisons, without their consent and often without their knowledge. If the Bill of Rights contains no guarantee that a citizen shall be secure against lethal poisons distributed either by private individuals or by public officials, it is surely only because our forefathers, despite their considerable wisdom and foresight, could conceive of no such problem....

CARL CARMER

TESTIMONY BEFORE THE FEDERAL POWER COMMISSION

IN THE MATTER OF CONSOLIDATED EDISON, WASHINGTON, D.C., MAY 1964

There is ominous talk along the Hudson today, talk from self-appointed ora-cles who declare that the march of cities is inevitable. In only a few years, we are told, the City of New York may stride like a wild irresistible monster in mile-by-mile steps ups the Great River of the Mountains. Soon the clamor of machines may be echoing from the Palisades—soon the noiseless deeps below the peaceful highlands may be sucked into the maw of the city.

Factories may line the flowing water, the white sails of countless sail-boats flutter no longer along the Hudson's reaches. The family motor-boats may decrease in numbers, the throbbing of countless engines may replace the silence of water pouring down from the Adirondack slopes of the north. We rejoice that private conservation societies have become our allies. Throughout the nation wherever the great pattern of rivers is attacked, wherever the Amer-ica of our forest-wise, mountain-wise, river-wise ancestors is threatened, the people have flocked to its defense. They speak out for the Allagash of Maine, the ever-loved Suwannee of Georgia and Florida, the Yukon of Alaska, the Yellowstone of California, the Potomac.

Our Federal Departments of Interior and Agriculture have published a report which begins: "The need to identify and preserve a nationwide system of free-flowing and undeveloped rivers, or segments of rivers, for their out-door recreation, scientific, historic, esthetic and symbolic values is urgent."

We believe that conservation is imperative today to the vast population of the United States. We believe that in no part of our great country is it more important than in the Hudson Valley.

We believe that true progress is made when the people preserve their

Excerpts from pages 990–93.

inheritance of scenic, historic, and recreational values as essential to their lives in work and in play along the Hudson. We believe that by alerting our citizens to the danger to their heritage, we may persuade them to assure progress and prevent regression. We believe that leaders of industry, if only they are informed, will work with us in seeking official remedies. Progress is a relative term and no more silly aphorism has been invented than that which declares it cannot be stopped. . . .

We believe that the people of our state will soon be welcoming to the Hudson's shores visitors from every country of the world. To them we would offer the same beauty our fathers offered travelers a century ago, the matchless loveliness of our stream, our valley, and our mountains. We would offer, too, the peace and healing our river gives, as it has always given, to those who seek its waters for respite from the tensions of their lives. We believe that the time for opposing those forces that would defile the Great River of the Mountains is now.

The people who live in this valley and love it want the Hudson kept as it is. A guide at the restoration of Boscobel wrote to the President of the United States about this not long ago: "I do not want to come out of the door of this beautiful house to face a massive commercial defacement of the other side of our river."

And a woman of historic, patriotic old Continental Village is reported as saying: "We have little left but beauty and now they would take that away from us."

Since I am convinced that all of America stands at a crossroads and that the Hudson, being what it is, should be the spear-point of a national drive against all agencies that would separate our people from their love of the American landscape as it has always been, I am committed to an uncompromising position. As a historian, I cannot find it in my heart to take any other stand.

PART 2

A DYING PLANET

IN THE LATE 1960S THE IMMEDIACY AND ENORMITY OF ENVIRON-
mental problems became increasingly clear to more and more Americans.
The documents collected in this section of the book, all dating from 1968 and
1969, contributed to the growing sense of crisis. They concern issues near and
far—from lead on the deteriorating walls of inner-city neighborhoods to starvation in India. They concern chronic problems (such as air pollution in Pittsburgh) and singular tragic events (such as the Santa Barbara oil spill). These
documents illustrate the diversity of people who turned their attention to the
environment at this moment; we hear from academics and union officers as
well as from a journalist and an elected official. Despite the variety of topics
addressed here, these documents illustrate a growing commonality of language, especially the language of crisis. Americans increasingly asked, What
can be done and by whom?

The first document is an excerpt from Paul R. Ehrlich's extremely influential book *The Population Bomb*. Ehrlich had been talking about the threats
of population growth for years, but this 1968 work brought him and the issue
international attention. Ehrlich is an ecologist and longtime professor of
biology at Stanford University. The urgency of his writing, felt so deeply by
many of his readers, came from a trip to India to conduct research on butter-
flies, a trip described briefly in chapter one of *The Population Bomb*, which is
excerpted here. Critics found his book alarmist, as they had Rachel Carson's
Silent Spring, but to many readers the crisis seemed real enough and the consequences of population growth were apparent—from salinization of irrigated
fields in the arid West to the scars of strip mining across Appalachia.

The second document is the cover of the second edition of *Whole Earth Catalog*, published in the spring of 1969. The brainchild of Stewart Brand, a lively counterculture figure in San Francisco, the catalog was a compendium of "tools" useful to the counterculture, to those who hoped to live less conventional lives, especially by doing for themselves things that the consumer culture had discouraged, including gardening. The cover may have been just as important as the contents of the catalog. The first edition, published in the fall of 1968, featured a photograph of the whole earth, as photographed from space. This image, and others like it, helped create a new, powerful metaphor: spaceship earth. The cover of the second edition, included here, featured the earthrise over the moon as photographed by the *Apollo* 8 mission. As a consequence of this image, and other images of the earth against the black backdrop of space, Americans increasingly thought of the limits to nature's bounty and the dire consequences should earth's natural systems fail.

The *Whole Earth* images spoke to the global nature of the environmental crisis, but most problems had immediate consequences for specific people. This was certainly the case for lead poisoning and environmental disease, which disproportionately affected poor minority communities in urban America. In 1969 journalist Jack Newfield published a poignant article describing this "epidemic of the slums," using the tragic story of Janet Scurry and her daughter to illustrate the extreme consequences of lead poisoning. Newfield helped ensure that lead poisoning would no longer be a "silent epidemic."

The next two documents in this section describe the deepening problems of air and water pollution in urban America. They also illustrate the activism undertaken by organized labor during the environmental crisis. Daniel W. Hannan, of the United Steelworkers of America in Clairton, Pennsylvania, testified before the Allegheny County Commissioners in the fall of 1969, hoping to secure tighter antismoke regulations and better enforcement. Hannan focused his testimony on the health consequences for workers and their families, the individuals who had to deal most directly with terrible air quality in and around steel and coke plants. Just a month after Hannan's testimony, members of the United Autoworkers in Detroit, led by Olga Madar, began to organize the Down River Anti-Pollution League to confront the terrible pollution problems along the River Rouge and in other industrial neighborhoods. In this organizing letter, Madar sounded hopeful that united action could bring relief and a healthier environment.

The last two documents in this section come from the West Coast—one generated by the Santa Barbara oil spill in 1969 and the other by the ongoing struggle to protect public health in such growing cities as Seattle. The first is the statement delivered to the Senate Subcommittee on Air and Water Pollution by Dr. N. K. Sanders, a professor of geography at the University of California–Santa Barbara, which was included in the record of the hearings on the proposed Water Pollution Control Act, conducted in the months following the Santa Barbara spill. Although this statement was not widely read, it described the commonly understood consequences of the oil spill. The statement reminds us of the importance of specific events in heightening the sense of crisis. The final document, from the Seattle–King County Department of Public Health, reminds us of the many ways local health officials worked to protect metropolitan residents. Like any good government report, the department's annual report for 1969 suggests that a corner had been turned, problems recognized, and solutions implemented. Pollution control, the document concludes, will be "the sound of the 70's."

PAUL R. EHRLICH

THE POPULATION BOMB

PROLOGUE

The battle to feed all of humanity is over. In the 1970s the world will undergo famines—hundreds of millions of people are going to starve to death in spite of any crash programs embarked upon now. At this late date nothing can prevent a substantial increase in the world death rate, although many lives could be saved through dramatic programs to "stretch" the carrying capacity of the earth by increasing food production. But these programs will only provide a stay of execution unless they are accompanied by determined and successful efforts at population control. Population control is the conscious regulation of the numbers of human beings to meet the needs, not just of individual families, but of society as a whole.

Nothing could be more misleading to our children than our present affluent society. They will inherit a totally different world, a world in which the standards, politics, and economics of the 1960s are dead. As the most powerful nation in the world today, *and its largest consumer,* the United States cannot stand isolated. We are today involved in the events leading to famine; tomorrow we may be destroyed by its consequences.

Our position requires that we take immediate action at home and promote effective action worldwide. We must have population control at home, hopefully through a system of incentives and penalties, but by compulsion if voluntary methods fail. We must use our political power to push other countries into programs which combine agricultural development and population control. And while this is being done, we must take action to reverse the deterioration of our environment before population pressure permanently ruins our planet. The birth rate must be brought into balance with the death rate

New York: Ballantine Books, 1968, excerpts from pages xi, 15–16, and 46–49. Copyright © 1968, 1971 by Paul R. Ehrlich. Used by permission of Ballantine Books, a division of Random House, Inc.

or mankind will breed itself into oblivion. We can no longer afford merely to treat the symptoms of the cancer of population growth; the cancer itself must be cut out. Population control is the only answer.

CHAPTER 1. THE PROBLEM

I have understood the population explosion intellectually for a long time. I came to understand it emotionally one stinking hot night in Delhi a couple of years ago. My wife and daughter and I were returning to our hotel in an ancient taxi. The seats were hopping with fleas. The only functional gear was third. As we crawled through the city, we entered a crowded slum area. The temperature was well over 100, and the air was a haze of dust and smoke. The streets seemed alive with people. People eating, people washing, people sleeping. People visiting, arguing, and screaming. People thrusting their hands through the taxi window, begging. People defecating and urinating. People clinging to buses. People herding animals. People, people, people, people. As we moved slowly through the mob, hand horn squawking, the dust, noise, heat, and cooking fires gave the scene a hellish aspect. Would we ever get to our hotel? All three of us were, frankly, frightened. It seemed that anything could happen—but, of course, nothing did. Old India hands will laugh at our reaction. We were just some overprivileged tourists, unaccustomed to the sights and sounds of India. Perhaps, but since that night I've known the *feel* of overpopulation.

A Dying Planet

Our problems would be much simpler if we needed only to consider the balance between food and population. But in the long view the progressive deterioration of our environment may cause more death and misery than any conceivable food-population gap. And it is just this factor, environmental deterioration, that is almost universally ignored by those most concerned with closing the food gap.

It is fair to say that the environment of every organism, human and nonhuman, on the face of the Earth has been influenced by the population explosion of *Homo sapiens*. As direct or indirect results of this explosion, some organisms, such as the passenger pigeon, are now extinct. Many others, such as the larger wild animals of all continents, have been greatly reduced in numbers. Still others, such as sewer rats and house flies, enjoy much enlarged populations. But those are obvious results and probably less important than more subtle changes in the complex web of life and in delicately balanced natural chemical cycles. Ecologists—those biologists who study the relationships of plants and animals with their environments—are especially concerned about these changes. They realize how easily disrupted are ecological systems (called ecosystems), and they are afraid of both the short- and long-range consequences for these ecosystems of many of mankind's activities.

Environmental changes connected with agriculture are often striking. For instance, in the United States we are paying a price for maintaining our high level of food production. Professor LaMonte Cole recently said, " . . . even our own young country is not immune to deterioration. We have lost many thousands of acres to erosion and gullying, and many thousands more to strip mining. It has been estimated that the agricultural value of Iowa farmland, which is about as good land as we have, is declining by 1% per year. In our irrigated lands of the West, there is the constant danger of salinization from rising water tables, while elsewhere from Long Island to Southern California, we have lowered water tables so greatly that in coastal regions salt water is seeping into the aquifers. Meanwhile, an estimated two thousand irrigation dams in the United States are now useless impoundments of silt, sand, and gravel."[1]

The history of similar deterioration in other parts of the world is clear for those who know how to read it. It stretches back to the cradles of civilization in the Middle East, where in many places deserts now occupy what were once rich and productive farmlands. In this area the process of destruction goes on today, still having, as in the past, ecologically incompetent use of water resources as a major feature. A good example is the building of dams on the Nile, preventing the deposit of nutrient-rich silt that used to accompany annual floods of the river. As almost anyone who remembers his high school geography could have predicted, the result has been a continuing decrease in the productivity of soils in the Nile Delta. As Cole puts it, "The new Aswan high dam is designed to bring another million acres of land under irrigation, and it may well prove to be the ultimate disaster for Egypt." The proposed damming of the Mekong could produce the same results for Vietnam and her neighbors.

Plans for increasing food production invariably involve large-scale efforts at environmental modification. These plans involve the "inputs" so beloved of the agricultural propagandist—especially fertilizers to enrich soils and pes-

---

1 Address delivered to the American Association for the Advancement of Science, December 27, 1967.

ticides to discourage our competitors. Growing more food also may involve the clearing of forests from additional land and the provision of irrigation water. There seems to be little hope that we will suddenly have an upsurge in the level of responsibility or ecological sophistication of persons concerned with increasing agricultural output. I predict that the rate of soil deterioration will accelerate as the food crisis intensifies. Ecology will be ignored more and more as things get tough. It is safe to assume that our use of synthetic pesticides, already massive, will increase. In spite of much publicity, the intimate relationship between pesticides on the one hand and environmental deterioration on the other is not often recognized. The relationship is well worth a close look.

WHOLE EARTH CATALOG

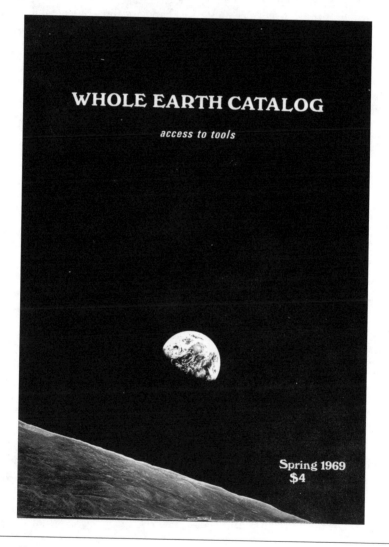

Cover (Spring 1969). Used with the permission of Stewart Brand.

JACK NEWFIELD

LEAD POISONING: SILENT EPIDEMIC IN THE SLUMS

VILLAGE VOICE (SEPTEMBER 18, 1969)

Except for its ironic name, Tiffany Street looks like a hundred other decaying streets in the Southeast Bronx. Mounds of uncollected garbage strewn all over. Idle young black and Puerto Rican men sitting on crumbling stoops. Mangy dogs looking through the garbage for scraps. Boarded-up, burnt-out houses freckled with graffiti.

At number 1051, on the fourth floor of a tenement whose dark halls stink from urine, Brenda Scurry was sitting in her clean, neat apartment telling me how her 23-month-old daughter Janet had died of lead poisoning in April.

Brenda, 23, is pretty, street smart, and black. She writes informed and angry letters to President Nixon and gets back impersonal form letters thanking her for interest in the "New Federalism." She has two other children, four- and five-year-old boys. Her husband works in the garment center, and sometimes lives at home. That morning Brenda had gone to the local public school and discovered her five-year-old sitting in a third-grade class and had to explain to the indifferent teacher that her child belonged in kindergarten.

"I used to live at 1113 Teller Avenue," she began quietly, but with a bitter-sweet edge to her voice. "Plaster from the walls started falling all over the place last November. I asked the landlord a couple times to do something about it, but he never did. Then in April one morning my daughter wouldn't eat anything. She started trembling and couldn't breathe. I got scared and she started to change colors. A neighbor called to a policeman and we took her to Morisannia Hospital. A doctor looked at her and told me to go home, that she would be okay. They didn't know what it was, but they sent me home. They asked me if Janet ever ate paint or plaster, and I told them yes. I went home but her temperature kept going up and down. After five days they gave her a blood

Excerpts used with the permission of *Village Voice*.

test for lead poisoning. And then she died the next day. The day after she died, the blood test came back positive. . . . Later they sent me a death certificate that said Janet died of natural causes. The doctors did an autopsy, but I still haven't got the results. I called the administrative director of the hospital twice, and they still haven't sent it to me. The hospital doesn't want to say it was lead, I guess."

"I asked welfare if they would pay for Janet's funeral, but they made me fill out a bunch of forms. So I paid for the funeral with the rent money. Then I asked welfare to pay for the rent. They said I had to fill out some other papers and that it would take a while. Then I got an eviction notice. I went to the central welfare office with it, but they still wouldn't give me any money. So I borrowed some money and moved out because I didn't want that landlord to put me on the street."

Lead poisoning is a disease endemic to the slums. The victims are hungry, unsupervised children between the ages of one and six, who get it by eating pieces of paint and plaster from flaking walls. Lead in paint was outlawed 20 years ago, but the bottom layers of walls in 800,000 run-down dwelling units in New York City still contain poisonous lead.

The city estimates that about 30,000 children each year suffer lead poisoning, but only 600 cases were reported during each of the last three years. (There are probably 300,000 victims nationally.) The early symptoms are vague nausea, lethargy, vomiting, crankiness, and doctors and nurses are not trained to look for it since they are told that lead has been outlawed as a paint ingredient. Ghetto parents are also ignorant of the disease. In three years, Harlem Hospital has not reported a single case of lead poisoning.

But surveys by researchers and activists keep discovering thousands of undiagnosed cases living in the ghettoes; lead poisoning has been called "the silent epidemic" by microbiologist Dr. René Dubos of Rockefeller University, a Pulitzer Prize winner last year.

According to doctors, 5 per cent of the children who eat lead die. Of those who survive, about 40 per cent suffer permanent brain damage, mental retardation, and deterioration of intelligence. A recent Chicago study of 425 children who had been treated for lead poisoning showed that 39 per cent had neurological disorders years later, and 22 per cent suffered from mental retardation as adults.

Two young, white, middle-class radicals have taken up the cause of lead poisoning, and for a year now have been waging a lonely crusade, bereft of money, manpower, or organizational support, to pressure the city and the health establishment and to alert parents. One is red-haired 30-year-old Paul Du Brul, the housing director of the University Settlement House on the Lower East Side. The other is bearded 24-year-old Glenn Paulson, co-chairman of the Scientists' Committee for Public Information.

Du Brul met me for lunch two weeks ago. He was particularly frustrated that day over the media's failure to take an interest in lead poisoning. He had called the [New York] Post's Joe Kahn earlier that morning, and Kahn had apologized, but the city desk wasn't interested in a story without a hard news peg. The [New York] Times had printed a story a few months before, but it had been buried in the real estate section on a Sunday, "where only landlords would see it."

Mrs. Scurry had written a personal letter to the Times man who wrote the Sunday story, telling him of her experience with her landlord, the hospital, and the Welfare Department. And he wrote back a moving letter how he felt lead poisoning was a tragic problem, but there just wasn't a news story in her case because the hospital denied lead was the cause of her daughter's death.

And lead was a story hard to make visible or dramatic for the television networks. It didn't involve famous leaders, or exotic militants, or public violence. How do you show a *process*, how do you show indifference, how do you show invisible, institutionalized injustice, in two minutes on Huntley-Brinkley? How do you induce the news department of a television network to get outraged about nameless black babies eating tenement paint, when the public health profession, school teachers, housing experts, scientists, the NAACP, and the politicians haven't given a damn?

Du Brul then began to explore some of the ramifications of this silent epidemic.

"Look," he said, "doctors say the effect of lead poisoning is to damage the nervous system. Kids can't concentrate. They become disruptive and lose points on I.Q. tests. So I think some of Jensen's findings (geneticist Arthur Jenson) about race and chromosomes might just be the effects of environmental conditions like lead poisoning. In the last ten years, 300,000 slum kids have been sent into the New York City public school system with lead poisoning. They're not culturally inferior; they're sick. What are now considered problems of remedial education might be doctor's problems, not teacher's problems."

Two nights later Du Brul and Paulson went to Judson Church on Washington Square to speak to a meeting of about 75 doctors, nurses, interns, and radical students sponsored by Health-PAC (Policy Advisory Center). Here the discussion focused on tactics, on how to fight the problems, how to make it visible.

Du Brul proposed a "fill the hospitals" strategy, coupled with the demand that every slum child between one and six receive a free laboratory test to determine if there is lead in his system. He said Chicago and Baltimore had been using a test that has proved 90 per cent effective, and the Lindsay Administration was holding back because of "bureaucratic bungling."

Some of the radical young doctors at the meeting disagreed. They said the hospitals would not, and could not, absorb the 30,000 walking cases in the city. They argued that such a tactic would collapse the already fragile hospital system. A few proposed rent strikes to force landlords to remove the paint or cover it up. Others suggested direct action at hospitals, particularly at Lincoln Hospital, which has lead on its own peeling walls. And still others urged a direct attack on the "slum system" as the root cause of lead poisoning. The meeting broke up at about 11 p.m., with even those few motivated on the issue divided over what to do first.

I spent the next few days working up an interior rage, trying to find out if anyone, with any responsibility, was doing anything about lead poisoning.

The NAACP had no program, nor any plans for any. The Department of Health, Education, and Welfare (the "good guys" in Nixonland) had no existing program and no funds allocated for any future program, but they did have a 21-page pamphlet. I could not find a copy of the pamphlet in any ghetto health office and had to acquire a copy from Glenn Paulson. The pamphlet turned out to be written in an opaque jargon that would hardly enlighten a second-year medical student, much less a welfare mother. The following is the second sentence on the first page: "Its etiology, pathogenesis, patho-physiology, and epidemiology are known."

Nineteen Congressmen, including William F. Ryan, have introduced a package of three bills to provide federal funds for a mass testing program in the slums. But the bills are given no chance of emerging from the limbo of committee, or even generating public hearings.

I also called the United Federation of Teachers to see if they were doing anything to detect cases in the schools, but no one called back.

Werner Kramarsky, Mayor Lindsay's staff man in the health field, advised me to call Health Commissioner Mary McLaughlin, to find out officially what the city was doing. When I called her, I was told the commissioner was not in, so I left a message. The next day the commissioner's press secretary called me and said I couldn't under any circumstances have a direct interview with the commissioner, but that he would answer any questions. I gave him a list of four.

After three days Dr. Felicia Oliver-Smith, the department's lead specialist, called back and reported that (1) the city had tested 7,000 children last year, compared to 35,000 in Chicago; (2) the city "hoped to" have a mobile testing unit "within one year"; (3) there were 725 cases last year, and more than 7,000 already so far this year; and (4) "We have no legal authority to make a landlord remove lead-based paint from tenement walls."

The only politician in the city who seems genuinely involved in the issue is Carter Burden, the Democratic-Liberal candidate for City Council in the polyglot East Harlem–Silk Stocking district. Burden has written angry letters to Lindsay, called press conferences, talked up the problem, and tried to energize grass-roots groups.

"It's terribly frustrating," Burden said last week. "The press just isn't interested at all. When I held a press conference in March on lead poisoning, not one daily paper and not one television station showed up. Just yesterday I had lunch with one of the religious leaders in East Harlem, and I tried to turn him on about lead. But he told me it was a phony issue, that *asthma* was a bigger community problem. . . . Kramarsky promised us three months ago there would be a crash program of 40,000 tests in the slums, but it didn't happen. Now I hear the city is about to start, but the summer is over, and 80 per cent of the cases develop during the summer. There was no reason for the delay."

Burden also revealed that a private blood test for lead poisoning had just been conducted in his district in the new, middle-income cooperative Franklin Plaza, which was completed *after* lead was banned from paint.

"That's a scandal," he said. "Of course there are no cases of lead poisoning in Franklin Plaza. But now they can release the results of the survey to prove lead is not a real problem in the slums." The survey was sponsored by Metropolitan Hospital and the American Cancer Society.

Some minimal testing program seems ready to be announced by the municipal bureaucracy. But it is not clear how the test will be distributed, or if it will reach the children who need it. And no one seems about to challenge the landlords to remove the deadly paint. And no one is doing anything about the hundreds of children now living in slum apartments with lead in the walls.

Today there is still lead in the walls of the apartment where Janet Scurry lived on Teller Avenue. There is lead in the two other tenements on the same block, owned by the same landlord who sent Mrs. Scurry the eviction notice.

There is still lead in the crumbling walls of Lincoln Hospital, where lead victims are sent. There are still thousands of undiagnosed, lead-poisoned children walking the streets, sitting in classrooms.

As Dr. René Dubos said at a conference on lead poisoning earlier this year: "The problem is so well defined, so neatly packaged with both causes and cures known, that if we don't eliminate this social crime, our society deserves all the disasters that have been forecast for it."

DANIEL W. HANNAN

TESTIMONY BEFORE THE ALLEGHENY
COUNTY COMMISSIONERS

SEPTEMBER 24, 1969

SUBJECT: Proposed new Allegheny County Smoke and Air Pollution Control
Rules and Regulations

GENTLEMEN:

My name is Daniel W. Hannan. I am president of Local Union 1557, United
Steelworkers of America, Clairton Works, Clairton, PA. I am privileged to
represent almost 4,000 members. My local union includes the members that
work in the largest by-product coke plant in the world. Thirty thousand tons
of coal are consumed every day, and we have monumental problems to match.
I am also an employee at the coke plant at Clairton Works as an instrument
repairman on the coke ovens. I don't pretend to be an expert on air pollution,
but I am here to tell it exactly how it is.

I appear before the distinguished commissioners of this county for a very
selfish reason. I am deeply concerned about the health, welfare, and well-being
of my members and their families both inside and outside the coke plant and
the citizens of this county. I am deeply concerned that we have documented
evidence that some members of our local union have been suffering from lung
cancer, black lung, silicosis, and other respiratory diseases. I am deeply concerned that the emission-level standards for particulate matter, sulfur oxides,
and other pollutants for the state and county must be set to match or even
better the federal criteria for air quality to protect the health of our members
and general public.

USWA Local 1557, USWA Collection, box 1, folder 23, "Pollution and Smog—1969–1972,"
Pennsylvania State University Special Collections.

I. W. Abel, President of our Steelworkers Union, has stated that air pollution in our environment can be most effectively controlled at the source. If coke-making facilities are judged to be an evident danger to the health of the community, how much more so are they a threat to the health of the workers within our plant? I hardly think that an aroused public and an alert government will stop at the achieving of clean air quality outside the plant when workers within are being subjected to higher concentrations of toxic elements than the general public ever was. Conditions in the coke plant demand immediate relief. My members almost constantly are exposed to this health and noise hazard. The impact of strong emission standards to control industrial pollution will have a positive impact upon the health of our members and the public as well. Occupational air quality of a plant must be made to compare favorably with the clean ambient air quality of the community.

One of the most distressing problems at the coke plant for me is the constant raping of the clean air of the atmosphere with little or no thought being given to the dire consequences of the health of the individuals that are exposed. This I say borders on sin. The environmental air pollution at the coke plant offers every conceivable type—odors, dust, noise, fumes, soot, gases, water, tars, oxides, and particulate matter—name it and we have it. I don't want to stand here and tell anyone we will settle for nothing less than clean mountain air. We know that this is impossible and air pollution to a degree is an inherent condition of a coke plant operation, but we know it is possible to significantly reduce this air garbage by strict operations, controls, and regulation. Several years ago with the charging of pulverized or powered coal into the coke oven, the situation has got progressively worse.

On April 30, 1969, Commissioners Forester and Hunt toured the Clairton coke plant and witnessed personally the working conditions our men endure at the coke ovens. Commissioner Forester was quoted as saying, "That was as close to hell as he has ever been." He also replied that no one could pay him enough to perform the duties that those men had to perform. A survey to evaluate employee exposure was conducted in May of this year by the industrial hygienists of the Pennsylvania Department of Occupational Health [that] seem[s] to bear out the remarks of the commissioner. The results of the tests (copy included) showed that those jobs greatly exceeded by many times the threshold limit value of 0.2 milligrams per cubic feet of air. The men must be living on borrowed time. Some time ago we had 26 members of our union overcome by deadly anhydrous ammonia vapors, and they were dispatched to the plant hospital after complaining of feeling nausea, weak, and ill. Just last week 6 members were overcome with large concentrations of sulfur dioxide and taken to the plant hospital for treatment. We hear of individual cases of members being overcome, but these are not generally reported to the local union. . . .

On behalf of our local union Air Pollution Committee, I would like to recommend that the present Allegheny County Air Pollution Advisory Committee be dismissed because it has become too large, apathetic to the need of stringent air pollution control, and industry-oriented. It includes members on the committee that are the worse offenders of air pollution. This is also true of state air pollution committees. We suggest a committee of ten or less, which should include public-spirited, interested citizens, men or women, doctors, scientists, and clergymen. We further suggest that Dr. Edward Stockton, Director of the Allegheny County Air Pollution Bureau, be retired because the committee feels that he has not served the best interests of the people of Allegheny County. Good strong emission standards should be determined by the breathers of the air in a community and not by those who have a vested interest in industrial facilities. . . .

One of our greatest natural resources is fresh, clean air. This life-sustaining element is God's free gift to mankind and no one in the name of progress and technological change has the right to alter this sacred gift. Our air pollution problem has to be solved and responsible people have to take the initiative to restore the air we breathe to its natural state. The alternative is just simply out of the question. Industry, government, and the general public must show its concern just as we are here today to show ours. Are you willing to saddle our children with a face mask as they wear eye glasses today? The type of air we breathe, Mr. Commissioners, is in your hands. Your life and my life may depend on the wisdom of the actions that you take in Allegheny County to reduce air pollution.

UNITED AUTO WORKERS

LETTER INITIATING THE DOWN RIVER ANTI-POLLUTION LEAGUE

NOVEMBER 25, 1969

Dear UAW Member and Spouse:

Do you remember this recent story reported by the news media: PLANES CANNOT LAND AT METROPOLITAN AIRPORT BECAUSE AIR IS SATURATED WITH SMOKE! Or this news story? ROUGE RIVER CATCHES ON FIRE, THREATENS OIL STORAGE TANKS AND I-75 BRIDGE!

Unbelievable? No! These reports were factual and serve as a grim reminder to each of us that the air and water on which we depend for life is rapidly being destroyed by continuing pollution. Not only has pollution in our streams, lakes, and rivers severely cut back our fishing, swimming, and boating activities, but the filth is so bad that even extensive chemical purification may be unable to provide us with reasonably clean drinking water!

Downriver residents have complained for some time now that their area suffers far more than others because pollution comes to their homes in large quantities every day. They have complained of various putrid odors brought to local air and water by industrial plants. Eyes and nose smart and burn because of particles and gases in the air, and scientists tell us our lungs and respiratory systems are damaged more each time we take a breath. Pollution also causes pitted or peeling paint on many Downriver homes and makes it difficult to keep our cars or clothing clean.

UAW Conservation and Recreation Departments Collection, box 2, folder "Down River Anti-Pollution League; Corres. & Reports, 1969," in the Walter P. Reuther Library at Wayne State University, Detroit.

We are convinced that these complaints are shared by a great many Downriver residents, who are concerned about the health and well-being of their families and annoyed to find that pollution is allowed to continue almost unabated despite the great increase in knowledge about it and laws and regulations that are supposed to prevent it.

We are also convinced that many share the feeling of one Downriver resident, that "each of us is frustrated because as individuals we feel helpless to do anything to stop pollution."

International Union's Executive Board has authorized the establishment of a special program to fight pollution in the Downriver area. River Rouge, Ecorse, Lincoln Park, and Wyandotte were selected for immediate action because in those communities the problem is most severe and complaints have been most numerous.

Your Union is confident that UAW members, joining with other citizens and organizations, can develop a plan of action to muster the united strength needed to effectively eliminate harmful pollution.

As a member of Local 174 who lives in this area, you and your family are urged to get together with others living nearby on Tuesday, December 9, 1969, for an evening which we hope will be devoted to getting more information, sharing views, and considering ways to fight pollution in your community. Your attendance will reflect your personal commitment and support for a healthier environment.

This meeting will be at 7:00 p.m. at the United Steelworkers #2659, John F. Kennedy Hall, 14024 Fort, Southgate. As refreshments will be served, would you please indicate on the postage-free return card how many of your family can attend and mail the card to us by December 5th.

Sincerely and fraternally,

Bard Young, Director
Region 1E, UAW

Phillip Terrana, President
Local Union #174, UAW

Olga M. Madar, Director
Department of Conservation and Resource Development

DR. N. K. SANDERS

"THE SANTA BARBARA OIL SPILL: IMPACT ON ENVIRONMENT" (1969)

The Union Oil spill has exerted, and will continue to exert, massive influence over the Santa Barbara environment. In very broad terms, the influence can be broken down into immediate and long-term categories. The most obvious immediate effect was the death of birds and mammals which normally lived near the sea surface: dolphins, seals, and the diving birds (grebes, cormorants, and loons). Less obvious was the filling of various intertidal and surficial life forms, usually small in physical size, which formerly inhabited the oil-covered beaches, rocks and kelp beds. The human community also showed a marked immediate response which may ultimately have far-reaching consequences.

Exact numbers of dead organisms resulting from the oil spill are impossible to gather. Over 1,500 birds have been brought to various agencies for oil removal (these have a 25% survival rate). Dead birds have not been collected and numbers could be at least as high as the oily survivors. One dead dolphin has washed up on the Goleta beach with an oil-blocked breathing hole; others have probably been affected. Two seals were brought to the Child's Estate cleaning station with oil-saturated fur and are being cared for there. These larger animals, seals and dolphins, will probably be able to survive longer in the oil-encrusted state and may not show up as deaths until later.

To those not personally, emotionally involved, these mammal and bird mortality numbers are not impressive. Almost that many motorists are killed every holiday weekend in the United States. Millions of intertidal life forms have been destroyed by the oil, but these are creatures not normally given much empathy by the average person. However, it is the death of these small

Statement delivered to the Senate Subcommittee on Air and Water Pollution of the Committee on Public Works, Water Pollution—1969, Part 4: Hearings on S. 7 and S. 544, 91st Cong., 1st sess., 1969, excerpt from pages 893–94.

forms that may have the greatest ecological significance.

Man, with his oil spill, has upset the ecological balance through destruction of the smaller organisms. Food chains have been broken which are needed for the continued existence of dolphins, seals, fish, lobsters and abalone. Eggs waiting to hatch have been destroyed. With 40 miles of the mainland intertidal zone and much of the Anacapa and Santa Cruz Island affected, aquatic life in the Santa Barbara Channel may become depleted. West Coast commercial fisherman, sportsmen and the consumer could feel the loss. Surface life forms living in the kelp have also been killed, which will reduce the productivity of that normally fertile habitat.

"Still," the realist is prone to think, "there are plenty of other places to set the nets, even if the Santa Barbara Channel is barren." Maybe so, but what of man's place in the chain? (After all, it's his environment, too.) The actual effect of the spill on the environment of man is impossible to quantify. To some, the smell is nauseous; others don't even notice. The spill may not greatly affect man physically, but he is profoundly touched emotionally.

Beach lovers and boat owners, a large percentage of the Santa Barbara population, have been deprived of their "rights" and are politically angry. The Chamber of Commerce fears lack of tourist revenue. Property owners are wondering about the beach frontage values. Even these intangibles are easier to quantify than the value a person places on being able to relax and reflect on a stretch of unpolluted coast.

These factors, subjective though they may be, show where the real impact of the oil spill will be felt. A community is largely a state of mind. It can be nothing more or less than what the local inhabitants make it. If oil continues to degrade the local environment, public apathy and disappointment replace optimism and pride. The blight will spread throughout the area and lead eventually to the production of yet another technological slum in a nation already over-endowed with wasted urban regions.

In summary, the effect of the oil spill on the environment is as follows: immediate effects [are the] death of a number of birds, sea mammals, and smaller life forms; future effects [are] possible depletion of some larger life forms in the Santa Barbara Channel, [and] probable reduction in the quality of human life.

INTRODUCTION, ANNUAL REPORT, 1969

On December 2, 1969, the City of Seattle observed its 100th anniversary as an incorporated municipality. In 1878, less than nine years after incorporation, Seattle hired its first health officer, Dr. F. W. Sparling. This marked the beginning of the Seattle Health Department that 72 years later merged with the King County Health Department on January 1, 1951, to form the Seattle–King County Health Department.

Before incorporation, Seattle's first physician, Dr. David S. Maynard, was a colorful character whose endeavors would fill a lexicon of job descriptions. Not the least of these was serving as director of the first hospital and previously as Indian Agent. As Indian Agent, he became the close friend of a tall, imposing Indian known as Sealth. In his youth, Sealth had been selected as chief on two occasions: first for his feats as a warrior, and later for his eloquence as a speaker. Chief Sealth and Dr. Maynard fished, hunted and traveled together. At times they talked for hours. For his friendship with the Indians, Dr. Maynard was castigated by some whites. However, when Washington's first governor, Isaac Stevens, made his territory-wide treaty-signing tour, he often encountered suspicion and resistance from the Indians. This was not the case in the Seattle area, and historians credit the smooth negotiations to the close relationship between Chief Sealth and Dr. Maynard.

When the time came for the pioneers to name their community, it was only natural that Dr. Maynard step forward to suggest the name of the old chief. The founding fathers agreed and the white man's pronunciation of the chief's name was applied to the new community. In this way Seattle's first doctor named Seattle.

As time passed, the old chief's visits to Seattle became less frequent. Toward the end he was a patient in the hospital run by his good friend, Dr. Maynard. The presence of Indians in the hospital disturbed some members of the white community. Ironically, Chief Sealth was not an entirely welcome

Excerpts from page 2.

recipient of the community's first form of public health services in a town that bore his name. . . .

When Dr. Maynard and Chief Sealth paddled up the Duwamish River to fish for salmon, the only sign of industry was a grey spiral of smoke rising above the mist, evidence that a cedar log was being charred so Indian carvers could chip out chunks to make a long canoe.

Today the mouth of the Duwamish is surrounded by, and filled in with, a vast complex of industrial plants. Here is the heaviest concentration of air pollution in the county. However, air pollution controls are being enforced: plants are installing control devices, and industrial emissions are diminishing because of enforcement by a multi-county agency that was spawned by the Health Department during the past decade.

As the 60s fade into the smog, we look to clearer horizons in the 70s. Suddenly, it seems, protest against pollution of the environment has become fashionable. The public has discovered the word "ecology." Right now the President of the United States and the Governor of Washington are preparing legislation to upgrade the quality of our environment.

Until recently environmentalists in public health have likened themselves to voices crying in the wilderness. Alarmed, the public is now joining ranks to prevent our environment from becoming a wilderness!

During the decade past public health people and people concerned with the public's health have taken positive steps to rehabilitate our environment, to prevent each of us from collectively poisoning ourselves with contamination of our air, water and soil. They have worked to protect our food and drink, our places of work, play and rest.

Medical research coupled with mass-immunization during the 60s has virtually eliminated widespread communicable disease in King County with the glaring exception of venereal disease.

Many of the major threats to the health of man are products of his own folly, related in varying degrees to his total environment. His body and brain are receptacles for the collective contamination of his own creation.

Sixty years ago, when Seattle had a population of 276,000, the commissioner of health, Dr. J. E. Chrichton, said, " . . . Laws must be made governing child labor, ventilation, slum life, workhouses, jails, and the crowding of people into great cities where they cannot live under proper environments."

Today, as 60 years ago, crowding of people into great cities endangers our environment. During the past decade laws were implemented to reduce environmental hazards. They include: (1) Burning garbage dumps were eliminated; improved methods of garbage and trash disposal were initiated. (2) Laws to control air pollution were promoted, passed and implemented; strong enforcement is conducted. (3) A water quality section was initiated, fulfilling a vital need. Strict surveillance has markedly increased the safety of drinking and bathing water throughout the entire county. (4) Sanitary sewer systems have been promoted and installed in the burgeoning suburbs, and the Metro sewage collection system has made Lake Washington the largest body of fresh water in the world to be rehabilitated. Before 1970 is over, no more Seattle sewage will pour into Puget Sound. (5) State regulations were adopted to prevent general use of the pesticides DDT and DDD. The Health Department's vector control personnel stopped using this material almost entirely at the beginning of the decade. (6) A specific food-borne illness prevention-detection program was started during the past decade, resulting in landmark activities in preventing salmonellosis, including preventing distribution of unpasteurized dry eggs by the Department of Agriculture. More has been done; more remains undone, including noise control. You'll be hearing about that. Pollution control, including noise, is the sound of the 70s.

PART 3

EARTH YEAR

THE DOCUMENTS COLLECTED IN THIS SECTION OF THE BOOK COME from 1970, the year of the first Earth Day and perhaps the peak year for American attention to environmental problems. These documents continue the theme of diversity—in the variety of environmental issues, the location of concern, and the types of people engaged—but they also express a nearly unified idea that now was the time for action.

The first document, the National Environmental Policy Act (NEPA), passed in 1969 but not effective until January 1, 1970, indicates the level to which Congress committed the federal government to reviewing its own impact on "all components of the natural environment." The act created the Council on Environmental Quality (in Title II, which is not included here), and it required that all federal agencies and entities receiving federal funding file an environmental impact statement before undertaking work. These impact statements have become a critical source of information for citizens concerned about changes in their communities and beyond. NEPA's inclusion in this collection serves to remind us of the critical role of the federal government in the search for environmental solutions as well as its role in creating environmental problems—by testing nuclear weapons, building interstate highways through neighborhoods and wilderness alike, spraying DDT and other pesticides, and so on.

Not everyone was keen on the expansion of government regulation in the interest of environmental protection, of course. In early 1970, William F. Buckley's *National Review*, an important outlet for increasingly influential conservative opinion, published an editorial focused on "the anti-pollution express."

Although several essays that appeared in the *National Review* acknowledged the real need for pollution control, this editorial took a sarcastic approach to the "escalating War on Pollution" and its consequences for escalating bureaucracy.

The third document in this section comes from South Carolina, where a number of residents of Beaufort County, home of Hilton Head Island, placed an ad in the *Columbia Record* as part of a successful campaign to keep the chemical giant BASF from building a massive plant along the coast. The ad smartly referenced the Cuyahoga River, which caught fire in June 1969—one of the many events that helped convince Americans that the environmental crisis was real. The Citizens Association of Beaufort County hoped to persuade the state legislature that South Carolina did not need to go down the same path as Ohio, that it might have "progress without pollution." The plant was never built, and Hilton Head continued to evolve into one of South Carolina's premiere vacation destinations.

Some conservatives remained wary of the environmental movement, especially given its push for increased government regulation, but President Richard Nixon understood the political value of supporting the environmental agenda. In February of 1970, Nixon delivered a lengthy "Special Message to the Congress on Environmental Quality," which is excerpted here. In the speech he described a thirty-seven-point program to help battle environmental problems at the federal level. Although Nixon avoided the language of crisis, the sheer breadth of his program speaks to his understanding of how much action needed to be taken. Five months after this speech, Nixon created the Environmental Protection Agency, an action that by itself secured his place among the most important environmental presidents.

The next document in this section is a *Seattle Post-Intelligencer* article from March 22, 1970, in which reporter Frank Herbert describes the goals of Native Americans who had occupied Fort Lawton, a U.S. Army base located on hundreds of acres along Puget Sound, north of Seattle's downtown. Here, filtered by the reporting process, we get a taste of the broadly shared Native American perspective on the environmental crisis, a perspective that tended to link nineteenth-century white imperialism over Native Americans with twentieth-century American imperialism over nature. The 1970 occupation had been sparked by the army's declaration that the property was surplus. Native Americans wanted to be sure that their ideas for that land's future would be heard. The demands of the United Indians, led by Puyallup chief Bob Satiacum, link the healing of Seattle's Native American community with the healing of the environment. The former fort is now Discovery Park, Seattle's largest park, and it contains the United Indians of All Tribes Daybreak Star Cultural Center, which provides many of the services Satiacum demanded.

Clearly, environmental activism led to some immediate successes, on both coasts and in the halls of government, but many environmental activists demanded more progress. Among them was the biologist and activist Barry Commoner, who thought Nixon's actions were insufficient to confront the "environmental crisis," a phrase that Commoner had popularized. In the late 1950s and early 1960s, Commoner had been an important voice in the battle to end nuclear testing and in the process he had become deeply suspicious of capitalism and technology. On the eve of Earth Day, Commoner gave a speech at Harvard University, excerpted in this collection, in which he argued, as he did many times, that population growth was not to blame for the environmental crisis. Instead, Commoner blamed "newly developed technology" that played a fundamental role in creating "a false prosperity"—something that in current parlance might be called unsustainability.

The first Earth Day created many significant documents, including the iconic *Pogo* poster included here, which proposes yet another explanation for the environmental crisis—individual behavior. Gaylord Nelson, the originator of the Earth Day idea and one of the leading congressional voices for the environment, also spoke to the power of individuals on Earth Day. In one of his many speeches during that week, Nelson advocated working with a "new urgency" to solve the nation's problems. He had a broad view of the crisis, addressing "rats in the ghetto" and "an environment of decency, quality, and mutual respect for all other human beings and all other living creatures." Nelson hoped to build a broad coalition for environmental action.

Even as Nelson made his plea in Denver, Nathan Hare—one of the founders of the journal *The Black Scholar* as well as the first Black Studies department in the nation at San Francisco State—wondered about the omission of black people from the ecology movement. In a special issue entitled "Black Cities: Colonies or City States," Hare describes the peculiar nature of black ecology, a consequence of the housing discrimination that kept urban blacks in overcrowded and polluted neighborhoods. In this scholarly essay, from which I have removed the footnotes, Hare describes yet another cause of the environmental crisis: racism.

Hare was certainly correct to point to racism's role in environmental inequality, but he may have underestimated the degree to which black communities participated in the environmental movement. The next documents—four letters written to Cleveland Mayor Carl Stokes as part of students' participation in Earth Day—reveal a remarkable range of concerns among the region's children. The first letter comes from the Daniel E. Morgan School, located in the Hough neighborhood. Just four years earlier, Hough had witnessed a deadly and damaging riot, part of the Long Hot Summer disturbances that rocked American cities in the second half of the decade. The list of "Recommendations for Controlling Litter" came from Mrs. LeGrande's science class, in which students brainstormed solutions to two overlapping problems in their neighborhood: trash and disorder. The second letter comes from Myron T. Herrick Jr. High, a south side school perched above the great steel mills of the Cuyahoga Valley.

The third letter, perhaps the most heartbreaking, comes from a Catholic school on the near West Side, near Ohio City, just above a train line that connected industry to the Cuyahoga. Together, these letters reveal an intimate connection between the urban crisis and the environmental crisis as well as the visceral nature of these crises for the children who lived through them. The fourth letter comes from a student who lived outside the city, in the middle-class community of Westlake. His letter reveals a genuine concern for the environment, but on a different level. These are just four of the hundreds of letters children wrote to Stokes, the first Black mayor of Cleveland, on or shortly after Earth Day.

The next two documents in this section also speak to the environmental crisis in Cleveland, providing some context for the Earth Day letters. In December 1970, Representative Louis Stokes spoke in favor of a river and harbors act that would have provided funding to clean up the Cuyahoga. In the process of supporting the bill, however, he further solidified the reputation of his city, and its river, as a terribly polluted place. Stokes, brother to the mayor, was an effective career politician, but here, in Congress, he sounded much like an environmental activist trained in ecology. Such was the success of the movement by the end of Earth Year. The other Cleveland-related document, a cartoon by *Plain Dealer* artist Ray Osrin, needs no explanation.

The next document in this section comes from the Portland area, where Eleanor Phinney and the Homeowners Preservation League worked to organize a conference on environmental issues. Phinney, who lived just south of Portland's expanding suburbs, supported urban planning as a means of shaping development and preserving the region's livable communities. In the years after Phinney wrote to the Oregon Environmental Council seeking support, Oregon became a leader in metropolitan planning. In 1979, Portland created its urban growth boundary as part of the struggle to contain suburban sprawl.

Meanwhile, in Pittsburgh, middle-class women worked to control another consequence of economic growth: air pollution. They founded the Group Against Smog and Pollution (GASP) in the fall of 1969. As the public service announcements included in this collection make clear, GASP blamed steel industry smokestacks for the pollution that continued to engulf their city. Just two days after Earth Day, Michelle Madoff, who lived in the upscale neighborhood of Squirrel Hill, wrote to WWSW Radio asking the public service director to read a series of announcements about her "reputable, hard-working volunteer organization." The announcements illustrate GASP's straightforward approach to pollution control: industry must spend the money and apply the technology that would end excessive emissions.

This long section ends with excerpts from the Clean Air Act, passed at the end of 1970. The act isn't particularly pleasant reading, but it does reveal how difficult air pollution regulation would be. Regulation would be dependent on science, and the science would be dependent on future research. The sheer complexity of the act also suggests the limits of the political system.

THE NATIONAL ENVIRONMENTAL POLICY ACT OF 1969

PUBLIC LAW 91–190 (JANUARY 1, 1970)

AN ACT

To establish a national policy for the environment, to provide for the establishment of a Council on Environmental Quality, and for other purposes.

Be it enacted by the Senate and House of Representatives of the United States of America in Congress assembled, that this Act may be cited as the "National Environmental Policy Act of 1969."

PURPOSE

SEC. 2. The purposes of this Act are: To declare a national policy which will encourage productive and enjoyable harmony between man and his environment; to promote efforts which will prevent or eliminate damage to the environment and biosphere and stimulate the health and welfare of man; to enrich the understanding of the ecological systems and natural resources important to the Nation; and to establish a Council on Environmental Quality.

TITLE I

Declaration of National Environmental Policy
SEC. 101. (a) The Congress, recognizing the profound impact of man's activity on the interrelations of all components of the national environment, particularly the profound influences of population growth, high-density urbanization, industrial expansion, resource exploitation, and new and

Excerpts from http://www.wilderness.net/NWPS/documents/publiclaws/PDF/91–190.pdf

expanding technological advances and recognizing further the critical importance of restoring and maintaining environmental quality to the overall welfare and development of man, declares that it is the continuing policy of the Federal Government, in cooperation with State and local governments, and other concerned public and private organizations, to use all practicable means and measures, including financial and technical assistance, in a manner calculated to foster and promote the general welfare, to create and maintain conditions under which man and nature can exist in productive harmony, and fulfill the social, economic, and other requirements of present and future generations of Americans.

(b) In order to carry out the policy set forth in this Act, it is the continuing responsibility of the Federal Government to use all practicable means, consistent with other essential considerations of national policy, to improve and coordinate Federal plans, functions, programs, and resources to the end that the Nation may—

    (1) fulfill the responsibilities of each generation as trustee of the environment for succeeding generations;

    (2) assure for all Americans safe, healthful, productive, and esthetically and culturally pleasing surroundings;

    (3) attain the widest range of beneficial uses of the environment without degradation, risk to health or safety, or other undesirable and unintended consequences;

    (4) preserve important historic, cultural, and natural aspects of our national heritage, and maintain, wherever possible, an environment which supports diversity and variety of individual choice;

    (5) achieve a balance between population and resource use which will permit high standards of living and a wide sharing of life's amenities; and

    (6 enhance the quality of renewable resources and approach the maximum attainable recycling of depletable resources.

(c) The Congress recognizes that each person should enjoy a healthful environment and that each person has a responsibility to contribute to the preservation and enhancement of the environment.

SEC. 102. The Congress authorizes and directs that, to the fullest extent possible: (1) the policies, regulations, and public laws of the United States shall be interpreted and administered in accordance with the policies set forth in this Act, and (2) all agencies of the Federal Government shall—

(A) utilize a systematic, interdisciplinary approach which will insure the integrated use of the natural and social sciences and the environmental design arts in planning and in decisionmaking which may have an impact on man's environment;

(B) identify and develop methods and procedures, in consultation with the Council on Environmental Quality established by title II of this Act, which will insure that presently unquantified environmental amenities and values may be given appropriate consideration in decisionmaking along with economic and technical considerations;

(C) include in every recommendation or report on proposals for legislation and other major Federal actions significantly affecting the quality of the human environment, a detailed statement by the responsible official on—

    (i)   the environmental impact of the proposed action,

    (ii) any adverse environmental effects which cannot be avoided should the proposal be implemented,

    (iii) alternatives to the proposed action,

    (iv) the relationship between local short-term uses of man's environment and the maintenance and enhancement of long-term productivity, and

    (v) any irreversible and irretrievable commitments of resources which would be involved in the proposed action should be implemented.

Prior to making any detailed statement, the responsible Federal official shall consult with and obtain the comments of any Federal agency which has jurisdiction by law or special expertise with respect to any environmental impact involved. Copies of such statement and the comments and views of the appropriate Federal, State, and local agencies, which are authorized to develop and enforce environmental standards, shall be made available to the President, the Council on Environmental Quality and to the public as provided by section 552 of title 5 United States Code, and shall accompany the proposal through the existing agency review processes;

(D) study, develop, and describe appropriate alternatives to recommended courses of action in any proposal which involves unresolved conflicts concerning alternative uses of available resources;

(E) recognize the worldwide and long-range character of environmental problems and, where consistent with the foreign policy of the United States, lend appropriate support to initiatives, resolutions, and programs designed to maximize international cooperation in anticipating and preventing a decline in the quality of mankind's world environment;

(F) make available to States, counties, municipalities, institutions, and individuals, advice and information useful in restoring, maintaining, and enhancing the quality of the environment;

(G) initiate and utilize ecological information in the planning and development of resource-oriented projects; and

(H) assist the Council on Environmental Quality established by title II of this Act.

SEC. 103. All agencies of the Federal Government shall review their present statutory authority, administrative regulations, and current policies and procedures for the purpose of determining whether there are any deficiencies or inconsistencies therein which prohibit full compliance with the purposes and provisions of this Act and shall propose to the President not later than July 1, 1971, such measures as may be necessary to bring their authority and policies into conformity with the intent, purposes, and procedures set forth in this Act.

SEC. 104. Nothing in Section 102 or 103 shall in any way affect the specific statutory obligations of any Federal agency (1) to comply with criteria or standards of environmental quality, (2) to coordinate or consult with any other Federal or State agency, or (3) to act, or refrain from acting contingent upon the recommendations or certification of any other Federal or State agency.

SEC. 105. The policies and goals set forth in this Act are supplementary to those set forth in existing authorizations of Federal agencies.

EDITORIAL

NATIONAL REVIEW BULLETIN (JANUARY 20, 1970)

"Pollution-control stocks, buoyed in part by visions of government spending as well as by 'concept' appeal, performed as the best group on a boom day with a clear sky" (*New York Times* report on the Jan. 2 market). So here we go again, full throttle on the anti-pollution express, shoveling in the billions to keep the fire hot and the pressure up. Don't get us wrong, mister. We're as strong for pure air, clean water, green trees and clean streets as the next man, and some of our best friends are ecologists. But really now, isn't there something pretty contrived, something rather nauseating not to speak of downright silly, in this currently escalating War on Pollution?

We've just got to find a new Cause every six months or so to keep the old adrenaline flowing. Fallout, civil rights, Vietnam, ghettos, hunger, poverty—they've lost their kicks and something fresh must be added. Anti-pollution has got everything it takes. The Sir Galahad devotion to purity. The St. Francis dedication to the well-being of all living things. The St. George assault on dragon Big Business. The mysterious links with choking, dread diseases, sterility, impotence. The always reliable thrill of the shadow of Doomsday. And it's a game anybody can play, Left to Right, liberal and conservative, student and banker.

So what are we going to do about pollution? Well, chum, we're going to have ourselves a ball, that's what we're going to do. We youngsters are going to have the grooviest demonstrations ever, and us senior citizens are going to forget our boredom what with the exciting petition campaigns and civic assemblies, to Save the Swamp, Ban Bangs, Deter Detergents, Suppress Sulphides, Annul Algae, Muzzle Monoxide, Clobber Corporations, and you name it. And

---

oh, those billions we're going to spend! Think of the grants for those new university courses in Bionomics and Neo-Ecology and Conservation Methodology and Symbiotic Equilibrium, and the new Departments of Environmental Science which, before you know it, will be pushing Black Studies off the top of the budget priorities. As for us politicians! No perfumes of Araby ever carried a more seductive scent to political noses than the garbage smells now rising. The Right is resolved this time not to let the Left get the jump. The President himself started the New Year by announcing a "now or never" war on pollution and signing a bill creating (and funding) an official Council on Environmental Quality. But Democratic Senators Henry Jackson and Edmund Muskie, sponsors of the bill, at once declared Mr. Nixon an imposter who was rolling in on the last freight. Governor Reagan made pollution the first topic of his New Year's message to the legislature; Governor Rockefeller raised him one by calling for a State Department of Environmental Conservation. The liberal politicians, like the liberal journalists, are panting with anticipation. In no time there will be more federal, state, municipal and volunteer civic commissions, commissioners, boards and committees, on pollution and environment than there are redwoods left in California. They'll all have jobs to allot and scads of grants and appropriations and contributions to spend all over the landscape.

And a funny thing. Not a single officer of this swelling anti-pollution army has suggested, or is likely to suggest, that one absolutely certain and no-cost way to start cleaning up the global environment would be to sweep up the junk and refuse right there in his own block.

"IS THIS WHAT YOU WANT FOR SOUTH CAROLINA'S WATERS?"

COLUMBIA RECORD (JANUARY 21, 1970)

"Neal Armstrong stepped out... of his spacecraft to become the first man to walk on the moon .... just a month after the Cuyahoga River in his native Ohio caught fire from oil pollution."

(AP wire service — Jan. 17, 1970)

Is this what you want for South Carolina's waters? BASF proposes to bring in 40,000 barrels of oil per day!

If YOU want to keep our rivers and marshlands free of oil pollution, join our fight. Fill in the coupon and mail to us — CABC, P. O. Box 5, Beaufort, S. C.

I want to help!

Name _____

Address _____

No. in family _____

CITIZENS ASSOCIATION OF BEAUFORT COUNTY

Rufus Taylor, President

PROGRESS without POLLUTION

Lottie D. Hamby Papers, South Carolina Political Collections, University of South Carolina.

RICHARD NIXON

SPECIAL MESSAGE TO THE CONGRESS
ON ENVIRONMENTAL QUALITY

FEBRUARY 10, 1970

To the Congress of the United States:

Like those in the last century who tilled a plot of land to exhaustion and then moved on to another, we in this century have too casually and too long abused our natural environment. The time has come when we can wait no longer to repair the damage already done, and to establish new criteria to guide us in the future.

The fight against pollution, however, is not a search for villains. For the most part, the damage done to our environment has not been the work of evil men, nor has it been the inevitable by-product either of advancing technology or of growing population. It results not so much from choices made, as from choices neglected; not from malign intention, but from failure to take into account the full consequences of our actions.

Quite inadvertently, by ignoring environmental costs we have given an economic advantage to the careless polluter over his more conscientious rival. While adopting laws prohibiting injury to person or property, we have freely allowed injury to our shared surroundings. Conditioned by an expanding frontier, we came only late to a recognition of how precious and how vulnerable our resources of land, water and air really are.

The tasks that need doing require money, resolve and ingenuity—and they are too big to be done by government alone. They call for fundamentally new philosophies of land, air and water use, for stricter regulation, for expanded

Excerpts from John T. Woolley and Gerhard Peters, *The American Presidency Project,* Santa Barbara, Calif., http://www.presidency.ucsb.edu/ws/index.php?pid=2757

government action, for greater citizen involvement, and for new programs to ensure that government, industry and individuals all are called on to do their share of the job and to pay their share of the cost.

Because the many aspects of environmental quality are closely interwoven, to consider each in isolation would be unwise. Therefore, I am today outlining a comprehensive, 37-point program, embracing 23 major legislative proposals and 14 new measures being taken by administrative action or Executive Order in five major categories:

— Water pollution control.
— Air pollution control.
— Solid waste management.
— Parklands and public recreation.
— Organizing for action.

As we deepen our understanding of complex ecological processes, as we improve our technologies and institutions and learn from experience, much more will be possible. But these 37 measures represent actions we can take *now*, and that can move us dramatically forward toward what has become an urgent common goal of all Americans: the rescue of our natural habitat as a place both habitable and hospitable to man. . . .

— I propose a Clean Waters Act with $4 billion to be authorized immediately, for Fiscal 1971, to cover the full Federal share of the total $10 billion cost on a matching fund basis. This would be allocated at a rate of $1 billion a year for the next four years, with a reassessment in 1973 of needs for 1975 and subsequent years. . . .
— I propose creation of a new Environmental Financing Authority, to ensure that every municipality in the country has an opportunity to sell its waste treatment plant construction bonds. . . .
— I propose that the present, rigid allocation formula be revised, so that special emphasis can be given to areas where facilities are most needed and where the greatest improvements in water quality will result. . . .
— Federally assisted treatment plants will be required to meet prescribed design, operation and maintenance standards, and to be operated only by State-certified operators.
— Municipalities receiving Federal assistance in constructing plants will be

required to impose reasonable users' fees on industrial users sufficient to meet the costs of treating industrial wastes.

— Development of comprehensive river basin plans will be required at an early date, to ensure that Federally assisted treatment plants will in fact contribute to effective clean-up of entire river basin systems. Collection of existing data on pollution sources and development of effluent inventories will permit systems approaches to pollution control.

— Wherever feasible, communities will be strongly encouraged to cooperate in the construction of large regional treatment facilities, which provide economies of scale and give more efficient and more thorough waste treatment. . . .

— I propose that State-Federal water quality standards be amended to impose precise effluent requirements on all industrial and municipal sources. These should be imposed on an expeditious timetable, with the limit for each based on a fair allocation of the total capacity of the waterway to absorb the user's particular kind of waste without becoming polluted.

— I propose that violation of established effluent requirements be considered sufficient cause for court action.

— I propose that the Secretary of the Interior be allowed to proceed more swiftly in his enforcement actions, and that he be given new legal weapons including subpoena and discovery power.

— I propose that failure to meet established water quality standards or implementation schedules be made subject to court-imposed fines of up to $10,000 per day.

— I propose that the Secretary of the Interior be authorized to seek immediate injunctive relief in emergency situations in which severe water pollution constitutes an imminent danger to health, or threatens irreversible damage to water quality.

— I propose that the Federal pollution-control program be extended to include all navigable waters, both inter- and intra-state, all interstate ground waters, the United States' portion of boundary waters, and waters of the Contiguous Zone.

— I propose that Federal operating grants to State pollution control enforcement agencies be tripled over the next five years—from $10 million now to $30 million in fiscal year 1975—to assist them in meeting the new responsibilities that stricter and expanded enforcement will place upon them. . . .

The Secretary of Health, Education and Welfare is today publishing a notice of new, considerably more stringent motor vehicle emission standards he intends to issue for 1973 and 1975 models—including control of nitrogen oxides by 1973 and of particulate emissions by 1975. . . .

— I propose legislation requiring that representative samples of actual production vehicles be tested throughout the model year. . . .
— I propose legislation authorizing the Secretary of Health, Education and Welfare to regulate fuel composition and additives. . . .
— I have ordered the start of an extensive Federal research and development program in unconventional vehicles, to be conducted under the general direction of the Council on Environmental Quality.
— As an incentive to private developers, I have ordered that the Federal Government should undertake the purchase of privately produced unconventional vehicles for testing and evaluation. . . .
— I propose that the Federal government establish nationwide air quality standards, with the States to prepare within one year abatement plans for meeting those standards. . . .
— I propose that designation of interstate air quality control regions continue at an accelerated rate, to provide a framework for establishing compatible abatement plans in interstate areas.
— I propose that the Federal government establish national emissions standards for facilities that emit pollutions extremely hazardous to health, and for selected classes of new facilities which could be major contributors to air pollution. . . .
— I propose that Federal authority to seek court action be extended to include both inter- and intrastate air pollution situations in which, because of local non-enforcement, air quality is below national standards, or in which emissions standards or implementation timetables are being violated.
— I propose that failure to meet established air quality standards or implementation schedules be made subject to court-imposed fines of up to $10,000 per day. . . .

I have ordered a re-direction of research under the Solid Waste Disposal Act to place greater emphasis on techniques for re-cycling materials, and on development and use of packaging and other materials which will degrade after use—that is, which will become temporary rather than permanent wastes. . . .

I have asked the Council on Environmental Quality to take the lead in producing a recommendation for a bounty payment or other system to promote the prompt scrapping of all junk automobiles. . . .

I have asked the Chairman of the Council on Environmental Quality to work with the Cabinet Committee on the Environment, and with appropriate industry and consumer representatives, toward development of such incentives and regulations for submission to the Congress. . . .

— I propose full funding in fiscal 1971 of the $327 million available through the Land and Water Conservation Fund for additional park and recreational facilities, with increased emphasis on locations that can be easily reached by the people in crowded urban areas.

— I propose that we adopt a new philosophy for the use of Federally-owned lands, treating them as a precious resource—like money itself—which should be made to serve the highest possible public good. . . .

— By Executive Order [11508], I am directing the heads of all Federal agencies and the Administrator of General Services to institute a review of all Federally-owned real properties that should be considered for other uses. This test will be whether a particular property's continued present use or another would better serve the public interest, considering both the agency's needs and the property's location. Special emphasis will be placed on identifying properties that could appropriately be converted to parks and recreation areas, or sold, so that proceeds can be made available to provide additional park and recreation lands.

— I am establishing a Property Review Board to review the GSA reports and recommend to me what properties should be converted or sold. This Board will consist of the Director of the Bureau of the Budget, the Chairman of the Council of Economic Advisers, the Chairman of the Council on Environmental Quality and the Administrator of General Services, plus others that I may designate.

— I propose legislation to establish, for the first time, a program for relocating Federal installations that occupy locations that could better be used for other purposes. . . .

— I also propose accompanying legislation to protect the Land and Water Conservation Fund, ensuring that its sources of income would be maintained and possibly increased for purchasing additional parkland. . . .

— I propose that the Department of the Interior be given authority to convey surplus real property to State and local governments for park and recreation purposes at a public benefit discount ranging up to 100 percent.

— I propose that Federal procedures be advised to encourage Federal agencies to make efficient use of real property. This revision should remove the budgetary penalty now imposed on agencies relinquishing one site and moving to another. . . .

— I propose that instead of simply paying each year to keep this land idle, we help local governments buy selected parcels of it to provide recreational facilities for use by the people of towns in rural areas. This program has been tried, but allowed to lapse; I propose that we revive and expand it.

— I propose that we also adopt a program of long-term contracts with private owners of idled farmland, providing for its reforestation and public use for such pursuits as hunting, fishing, hiking and picnicking. . . .

I have ordered that the United States Patent Office give special priority to the processing of applications for patents which could aid in curbing environmental abuses. . . .

I will shortly ask a group of the nation's principal industrial leaders to join me in establishing a National Industrial Pollution Control Council. . . .

FRANK HERBERT

HOW INDIANS WOULD USE FORT

SEATTLE POST-INTELLIGENCER (MARCH 22, 1970)

The first thing the United Indians want to do with Ft. Lawton is set up an "environmental preserve" which will help teach whites "how to stop destroying the earth."

The next thing they want to do is change the place's name.

These priorities and others were explained by Bob Satiacum, blood chief of the Puyallups and leader of the "Fort Lawton Re-occupation" in an interview with the *P-I.*

Satiacum and his followers also want to establish on this land an Indian adoption agency where orphaned Indian children would be given to Indian families.

High on their priority list for the site is an "Indian Half-Way House" where Indians coming off reservations can be prepared for "going outside."

The Number One item, though, is the "environmental preserve."

Indians will meet today to decide whether to remain encamped outside the base as they have been since March 10 or leave the area. Meanwhile the Army has strung barbed wire on the beaches below the fort to thwart a repeat of last Sunday's invasion.

The United Indians want the 1,100 acres restored to a state they describe as "Indian natural." Satiacum explains:

"We want the land to be self-sustaining and not burdened to destruction by the population it supports."

As to changing the name:

"We don't think it right a place designed to teach us survival should be named for a man who ignored international boundaries (Mexico's) to put a great Indian leader (Geronimo) in bondage."

Used with permission.

This refers to the raid which Lawton led into Mexico to capture Geronimo and return him to custody in the United States.

Satiacum said he found it "laughable" that Lawton's raid into Mexico gets "so much historical applause" in the U.S. while Indian attempts "to reoccupy our own land draw down the wrath of the courts."

United Indians' leaders are particularly insistent that the word "Fort" be removed from this land along Puget Sound.

He described himself as "outraged" that some whites object to Indian claims on this land because "Indians don't know how to use the land." He said:

"We don't think we have to destroy the earth to prove or 'improve' its value. It would take a pretty insensitive human being to ignore today's evidence that our environment is being made deadly to all of us."

Does he really think Indians can regain this land?

"That's why we're here."

Regaining it, does he think he can teach the surrounding culture anything?

"We must or we're all doomed."

Satiacum is quick to add that the United Indians make no claim to "all the land taken from us in this area." He says:

"We're practical about this. We don't want to dispossess people. We do have a moral right to this particular land, though. It was taken from us by treachery, by germ warfare and by force of arms. We were defeated.

"In the process, though, our conquerors are defeating themselves. What horrified us is that they are killing us all in the process. We must save this land."

Physically, what would this 1,100 acres of land look like under Indian occupation?

"Most of the land would be restored to its most supportive natural state. There would be berries, trees, game. Beach life would be allowed to restore itself. Somewhere among the trees, we would build a true Indian long house to be used as school and museum.

"For the school, we want to show how proud we are of our own culture and heritage. We want to pass that along—not to just our own descendants, but to whites, too."

Satiacum is particularly interested in creating a new library of Indian lore on tape, making it available at the "long house school."

He is concerned about "inaccuracies" put into the dominant educational system by white-oriented sociologists, anthropologists, and historians. He said:

"Many times these investigators heard only what we wanted them to hear. We were afraid to tell them too much about ourselves. Such knowledge was used to destroy us.

"Our old people and the fortunate few they taught have a treasure house of knowledge, a tremendous store of lore. This is knowledge about how to restore our land and save it from destruction.

"We must recapture that lore before it dies with the last few who hold it in trust. This excess federal land which was once ours—and is morally ours now—would make an ideal place to begin that restoration."

BARRY COMMONER

HARVARD UNIVERSITY LECTURE

DELIVERED ON APRIL 21, 1970

Everyone now knows that the environmental crisis is upon us. We no longer need to convince ourselves that we are poisoning our waters, destroying our soil, and clogging the air with pollutants. We also know that pollution is much more than a nuisance or threat to health; it is a symptom of a disastrous incompatibility between the environmental system that supports us and the way in which we act on it. If we fail to rectify this suicidal relationship between ourselves and the world in which we live, we cannot last very long on this planet which is our habitat.

All of this most people now know. What is not so clear is how we got into this mess and what we need to do to get out of it. I want to direct my comments to those questions.

Answers have already been provided and it is useful to examine some of them. In last Sunday's *New York Times* there was a full-page ad which asserts that we cannot cure our environmental ills unless we first control the rising U.S. population. This view—that pollution is due to too many people—has the backing not only of those who think there are too many people in the U.S. for reasons of their own, but also has the scientific authority of the President of the National Academy of Sciences behind it. What truth can we find in these assertions?

In my opinion, population trends in the U.S. cannot be blamed for the deteriorated condition of the environment. A few relevant statistics tell the story. Nearly every one of the pollutants that now concern us—artificial radioactivity, detergents, chemical fertilizers, smog, nitrogen oxides, insecticides—made their appearance in our environment only 20–25 years ago. The

Excerpts from the Barry Commoner Papers, Box 35, Manuscript Division, Library of Congress, Washington, D.C.

environmental crisis is the result of the rapid burgeoning of the burden of these pollutants that has been imposed on the environment. In most cases the increases in the last 20–25 years have been of the order of 500 to 1,000 percent. The record is perhaps held by inorganic nitrogen fertilizers; the amount of this material imposed on the U.S. soil since 1945 has increased 14-fold. In other words, the stress on the environment that now concerns us has increased by an order of magnitude in the last 20–25 years.

What about the concurrent change in the population? This figure has only risen by 40–45% in that time. In effect, then, the intensification of pollution is ten times too large to be accounted for by the increase in our population. Of course, if there were no people in the country, there would be no pollution problem, but the fact of the matter is that there simply has not been a sufficient rise in the U.S. population to account for the enormous increase in pollution levels. Those people who believe that the country is too crowded will have to find some other reason to explain their position, and it is a serious mistake to becloud the pollution issue with the population problem for the facts will not support it. There is a relationship between the general industrial situation in the U.S. and other developed countries and the authentic population crisis which is developing in the tropical countries of the world. That is a matter that I will return to shortly.

Another argument is that pollution is the necessary accompaniment of our rising standard of living and that the cure is, to quote some of the eco-activists, to "reduce consumption." The favorite statistic is that the U.S. contains 6–7% of the world population but consumes more than half the world's resources and is responsible for that fraction of the total environmental pollution. But this statistic hides another vital fact: that not everyone in the U.S. is so affluent. For that reason the simple test of the value of the slogan "Consume Less" as a basis for social action on the environment would be to tell it to the blacks in the ghetto. The message will not be well received, for there are many people in this country who consume less than is needed to sustain a decent life.

A few facts to help out. For example, if we use the GNP as a measure of our overall standard of living, then we discover that during the period in which pollutant production rose by an order of magnitude, the GNP increased only by about 250%. In other words, even disregarding the inequities in the distribution of the country's wealth there is a four-fold discrepancy approximately between the overall increase in our affluence and the overall increase in environmental pollution. Again, this factor fails to explain the actual increase in pollution level.

What is the reason for our trouble then? In my opinion environmental pollution is the consequence of the kind of technology that we have used to enhance our productivity in the last 20–25 years. During this time there have been enormous changes in the way in which we carry out our industry and agriculture. New types of power plants have been built, automobile engines have increased in power, new agricultural chemicals have been massively introduced, soap has been replaced by synthetic detergents, aluminum has replaced the steel beer cans. The tragic fact is that in not one of these instances has the newly developed technology taken into account its effects on the environment. When the new detergents were developed, the chemists and engineers were careful to make their product conform to the best chemical and engineering practice. Careful research was conducted to make sure that the detergents would wash well and sell well. What they failed to do was to make an inquiry of the acceptability of this product to the ultimate consumer—the bacteria of decay on which we must rely to break down any organic matter introduced into the environment. It turned out that the detergents were not acceptable to them, and they accumulated in massive mounds of foam on our water ways. The trouble with detergents is not that they have improved our standard of living, but that they have done so at the expense of a fatal environmental error.

In the same way the agronomists who have fostered the massive use of inorganic nitrogen fertilizer were able to improve crop yields and the farmer's financial position. But they did so at the expense of introducing sufficient nitrate into the environment to cause massive pollution in surface and ground waters in nearly every agricultural area. Surely the engineers in Detroit have improved—if you want to call it that—the quality of life by giving so many people so many cars. But what they failed to know was that these same cars and their high-powered, hot-running engines would produce sufficient nitrogen oxides to trigger off the massive accumulation of smog that pollutes every city in the country. And, having built cars which soon needed to be junked, they also added to the huge burden of solid waste that engulfs us.

In other words, we are living in a false prosperity. Our burgeoning industry and agriculture has produced lots of food, many cars, huge amounts of power, and fancy new chemical substitutes. But for all these goods we have been paying a hidden price: the potentially fatal destruction of the very environmental system that not only supports us but the industries themselves. What this tells us is that our system of productivity is at the heart of the environmental problem. It is designed in such a way as to rely for its economic success on the evasion of its true costs in environmental destruction. If power companies were required to show on electric bills the true cost of power to the consumer, they would have to include the extra laundry bills resulting from soot, the extra doctor bills resulting from emphysema, the extra maintenance bills due to erosion of buildings. Their true account books are not in balance, and the deficit is being paid by the lives of the present population and the safety of future generations. This tells us that despite the claims of the populationists and those who would have us reduce our consumption, the environmental crisis is not to be explained by the simple accumulation of people or their increased appetites for goods. Instead, there is something wrong in the way in which the goods that we consume have been produced, and this fault is so massive as to raise serious questions about the viability of our economic system if for the sake of survival we are required to prevent its environmental costs. For the basic issue which the environmental crisis raises is whether or not our system of productivity can remain unchanged and yet meet the demands of compatibility with the environment. . . .

WALT KELLY

POGO POSTER: "WE HAVE MET THE ENEMY, AND HE IS US"

EARTH DAY 1970

84

GAYLORD NELSON

EARTH DAY SPEECH

I congratulate you, who by your presence here today demonstrate your concern and commitment to an issue that is more than just a matter of survival. *How* we survive is the critical question.

Earth Day is dramatic evidence of a broad new national concern that cuts across generations and ideologies. It may be symbolic of a new communication between young and old about our values and priorities.

Take advantage of this broad new agreement. Don't drop out of it. Pull together a new national coalition whose objective is to put Gross National Quality on a par with Gross National Product.

Campaign nationwide to elect an "Ecology Congress" as the 92nd Congress—a Congress that will build bridges between our citizens and between man and nature's systems, instead of building more highways and dams and new weapons systems that escalate the arms race.

Earth Day can—and it must—lend a new urgency and a new support to solving the problems that still threaten to tear the fabric of this society . . . the problems of race, of war, of poverty, of modern-day institutions.

Ecology is a big science, a big concept—not a copout. It is concerned with the total ecosystem—not just with how we dispose of our tin cans, bottles and sewage.

Environment is all of America and its problems. It is rats in the ghetto. It is a hungry child in a land of affluence. It is housing that is not worthy of the name; neighborhoods not fit to inhabit.

Environment is a problem perpetuated by the expenditure of $17 billion a year on the Vietnam War, instead of on our decaying, crowded, congested, polluted urban areas that are inhuman traps for millions of people.

Delivered on April 22, 1970, in Denver, Colorado, partial text as found in the Gaylord Nelson Collection at the Wisconsin Historical Society, http://www.nelsonearthday.net/collection/422-denverspeech.htm

If our cities don't work, America won't work. And the battle to save them and end the divisiveness that still splits this country won't be won in Vietnam.

Winning the environmental war is a whole lot tougher challenge by far than winning any other war in the history of Man. It will take $20 to $25 billion more a year in Federal money than we are spending or asking for now.

Our goal is not just an environment of clean air and water and scenic beauty. The objective is an environment of decency, quality and mutual respect for all other human beings and all other living creatures.

Our goal is a new American ethic that sets new standards of progress, emphasizing human dignity and well-being rather than an endless parade of technology that produces more gadgets, more waste, more pollution.

Are we able to meet the challenge? Yes. WE have the technology and the resources.

Are we willing? That is the unanswered question.

Establishing quality on a par with quantity is going to require new national policies that quite frankly will interfere with what many have considered their right to use and abuse the air, the water, the land, just because that is what we have always done.

NATHAN HARE

BLACK ECOLOGY

THE BLACK SCHOLAR (APRIL 1970)

The emergence of the concept of ecology in American life is potentially of momentous relevance to the ultimate liberation of black people. Yet blacks and their environmental interests have been so blatantly omitted that blacks and the ecology movement currently stand in contradiction to each other.

The legitimacy of the concept of black ecology accrues from the fact that: (1) the black and white environments not only differ in degree but in nature as well; (2) the causes and solutions to ecological problems are fundamentally different in the suburbs and ghetto (both of which human ecologists regard as "natural [or ecological] areas"; and (3) the solutions set forth for the "ecological crisis" are reformist and evasive of the social and political revolution which black environmental correction demands. . . .

With the industrialization and urbanization of American society, there arose a relatively more rapid and drastic shift of blacks from Southern farms to Northern factories, particularly during periods when they were needed in war industries. Moreover, urban blacks have been increasingly imprisoned in the physical and social decay in the hearts of major central cities, an imprisonment which most emphatically seems doomed to continue. At the same time whites have fled to the suburbs and the exurbs, separating more and more the black and white worlds. The "ecology crisis" arose when the white bourgeoisie, who have seemed to regard the presence of blacks as a kind of pollution, discovered that a sample of what they and their rulers had done to the ghetto would follow them to the suburb. . . .

In addition to a harsher degree of industrial pollutants such as "smoke, soot, dust, fly ash, fumes, gases, stench, and carbon monoxide"—which, as in the black ghetto, "there is no wind or if breezes are blocked dispersal will

---

Excerpts from pages 2–8. Used with permission.

not be adequate"—the black ghetto contains a heavier preponderance or ratio, for instance, of rats and cockroaches. These creatures comprise an annoyance and "carry filth on their legs and bodies and may spread disease by polluting food. They destroy food and damage fabrics and book bindings." Blacks also are more exposed to accidents, the number four killer overall and number one in terms of working years lost by a community. . . .

At the heart of this predicament, though not alone, is the crowded conditions under which most black persons must live. Black spatial location and distribution not only expose blacks to more devastating and divergent environmental handicaps; they also affect black social and psychological adjustment in a number of subtle ways. . . .

The social and psychological consequences of overcrowding are tangled and myriad in degree. To begin with, the more persons per unit of space, the less important each individual there; also, the noisier the place, other things equal, and the greater the probability of interpersonal conflict. Studies show that there is a greater hearing loss with age and that much of it is due to honking horns, loud engines and general traffic noise. The importance of space to contentment also is suggested by the fact that in a survey of reasons for moving to the urban fringe, that of "less congested, more room" was twenty times more frequently given than the fact that the environment was "cleaner". . . .

But the residential pollution of blacks rests not alone in overcrowding and the greater prevalence of unsightly and unsanitary debris and commercial units such as factories. The very housing afforded blacks is polluted. This fact is crucial when we consider that the word "ecology" was derived by a German biologist from the word "aikos" meaning "house." A house, like the clothes we wear, is an extension of one's self. It may affect "privacy, child-rearing practices, and housekeeping or study habits." Three of every ten dwellings inhabited by black families are dilapidated or without hot water, toilet or bath. Many more are clearly fire hazards.

The shortage of adequate housing and money for rent produces high rates of black mobility which have far-reaching effects on the black social environment. It means that blacks will disproportionately live among strangers for longer periods of time and, in the case of children, attend school in strange classrooms. . . .

No solution to the ecology crisis can come without a fundamental change in the economics of America, particularly with reference to blacks. Although some of the ecological differentials between blacks and whites spring directly from racism and hence defy economic correlations, many aspects of the black environmental condition are associated with basic economics. Blacks are employed in the most undesirable or polluted occupations, lagging far behind their educational attainment. About two-thirds work in unskilled and semiskilled industries. . . .

Thus the reformist solutions tendered by the current ecology movement emerge as somewhat ludicrous from the black perspective. For instance, automobiles are generally regarded to be the major source of air pollution. This is compounded in the case of blacks by the relatively smaller space in which they must live and drive amid traffic congestion and junked cars. On top of this, white commuters from the suburbs and the outer limits of the central city drive into the central city for work or recreation and social contacts, polluting the black environment further. In every region of the country there has been a direct parallel between the increase in the number of cars and the growth of the suburban and fringe population. Although automobile manufacturers are the chief profiteers, the contradiction of alien automobile polluters who daily invade and "foul the nest" of black urban residents remains.

Some of these commuters are absentee landlords who prevail as "ghetto litterbugs" by way of corrupt and negligent housing practices. Thus blacks suffer the predicament wherein the colonizer milks dry the resources and labor of the colonized to develop and improve his own habitat while leaving that of the colonized starkly "underdeveloped". . . .

The real solution to the environmental crisis is the decolonization of the black race. Blacks in the United States number more than 25,000,000 people, comprising a kidnapped and captive nation surpassed in size by only twenty other nations in the entire world. It is necessary for blacks to achieve self-determination, acquiring a full black government and a multi-billion-dollar budget so that blacks can better solve the more serious environmental crises of blacks. To do so, blacks must challenge and confront the very foundations of American society. In so doing, we shall correct that majority which appears to believe that the solution lies in decorating the earth's landscape and in shooting at the moon.

LETTERS FROM SCHOOLCHILDREN TO CARL STOKES, MAYOR OF CLEVELAND

EARTH DAY 1970

(1)

April 22, 1970

Recommendations for Controlling Littering
1. All trash should be disposed of in suitable containers.
2. The city should provide trash cans at every busy corner of the city.
3. Garbage collectors should be more careful when collecting garbage.
4. Every car owner should be required to have a litter bag in his vehicle, available to occupants.
5. Garbage should be collected twice weekly in heavily populated areas.
6. Street clubs should concentrate more on the problems of pollution.
7. All boys could volunteer to help clear their neighborhood of litter.
8. Dog owners should be required to walk their dogs instead of allowing them to run free in the neighborhood.
9. Breaking bottles and glass on the street and sidewalks should be outlawed.
10. The city should do more to rid the neighborhoods of rats and other pests.
11. The city should provide supervisors for the Morgan Playfield.
12. All laws (present and new) must be enforced.

> The Sixth Grade Boys and Girls of
> Daniel E. Morgan School (Science)

Carl Stokes Papers, container 75, folder 1438, at Western Reserve Historical Society.

(2)

Cleveland, Ohio
Myron T. Herrick Jr. High
McBride Ave. 5407
April 29, 1970

Dear Mayor Stokes,

Where, when, and how are you going to clean the air and water of Cleveland? We Clevelanders need it very bad. We found that each day and year the water pollution is getting bad for us to live.

Air pollution is bad for young and old. Mainly the pollution is coming from our factories and mills. Smoke from the smoke stacks from the factories causes smog and smell.

Water pollution is bad for young and old too! The people are dumping garbage in the water. The factories are dumping substances in the water that causes a disease and kills the fish.

People pollution is awful. Because the earth is overcrowded and it keeps building up.

Where! Cleveland

When! Now

How! More garbage cans. Put a device on the smoke stacks or chimney. Reduce population.

Thank you for your trouble,

Joyce Bowman
Tina Palady
Radojka Pavlovic
Students 7th grade

(3)

2259 Columbus Rd.
Cleveland 13, OH
Earth Day, 1970

Mr. Stokes:

I am writing to you because I and others have a problem which I think you can help us solve. Willy Hill on Willy Street is a dump. It has garbage and filth all around it. It involves danger. There are also rats down there. I am writing to you, so that you will please have someone to come and clean the mess. It really doesn't give a good impression of St. Wendelin School to which I go.

If you have it cleaned, I promise I will do what I can to keep it that way. It should look like God wanted it to be. Thank you.

Melissa Stevens

(4)

23903 Westwood Rd.
Westlake, Ohio 44145
April 22, 1970

Mayor Carl B. Stokes
City Hall
Cleveland, Ohio 44101

Dear Mr. Mayor:

This letter is concerning the pollution problem of Cleveland. I realize that as Mayor, you're not responsible for everything that goes on in the NOW CITY. But you could pressure some (if not all) of the growing industries in Cleveland, like Republic Steel, Muny Light Plant and Edgewater Park. These are just some examples but they're major concerns.

Another real good example is the "good ol" Cuyahoga River. I mean not every city can say they have a river that catches on fire. And that's nothing to be proud of.

On a recent school field trip to the Cleveland Health Museum, I was watching out the window of the bus. As we reached downtown, I looked out at the sky. It was so full of waste products, I could hardly see the top of the Terminal Tower. I thought that was pretty disgusting.

Can you tell me what you, as Mayor, are doing to solve this rapidly increasing problem? This is Environment Week, and in school we are discussing these various problems and different students are writing letters to different people concerning this crisis.

I chose to write to you, not because I have to for an assignment, but because you should know more about it than I. If my letter is answered, I will tell the other concerned students and teachers in my Biology and English classes what you are doing to *HELP*! save Cleveland. Thank you.

Sincerely yours,

Robert L. Tasse

ADDRESS IN CONGRESS SUPPORTING RIVERS AND HARBORS AND FLOOD CONTROL ACT OF 1970

DELIVERED ON DECEMBER 7, 1970

Mr. Speaker, I rise in support of the bill. It is well known that the Cuyahoga River, which runs through my congressional district, is one of the most polluted streams in the United States. It will live in infamy as the only river in the world to be proclaimed a fire hazard. In fact, in November of last year the British Broadcasting Co. called the Greater Cleveland area the pollution capital of the world.

Section 108 of this bill will go far in alleviating this overwhelming problem. It was taken from a bill I introduced in September—H.R. 19091—and authorizes a massive cleanup of the Cuyahoga. I am very grateful to both the distinguished chairman of the Public Works Committee, Mr. FALLON, and the distinguished subcommittee chairman, Mr. BLATNIK, for their support of my bill. I also offer my sincere thanks to the gentleman from New Jersey (Mr. HOWARD) for the leadership role he played in moving the measure along so rapidly and to my Ohio colleague, the distinguished ranking minority member, Mr. HARSHA, for his work in broadening the scope of the bill to include the entire Cuyahoga Basin.

Mr. Speaker, perhaps if my colleagues could learn of the enormous ecological problems associated with the Cuyahoga, they could share the committee's sense of urgency in getting this project off the ground.

Today, 17 streams and over 500 outfalls of sewage mixed [with] industrial wastes and storm and combined sewer overflows discharge into the Cuyahoga along its 85-mile length from its origin at Lake Rockwell, through the industrial giants of Akron and Cleveland, to its mouth in downtown Cleveland at Lake Erie.

Congressional Record, v. 115, part 14 (91st Congress, 1st Session), page 40150.

During the past few years, there have been efforts made at the local level by both public agencies and private organizations to eradicate this stigma from our landscape. These efforts, although fruitful, could not be of sufficient magnitude to accomplish this goal. We can be encouraged, however, by the fact that such cleanup programs can result in improvements in receiving water quality. It is imperative, therefore, that we continue an aggressive program for water pollution control in the Cuyahoga River. Further, I believe that if we demonstrate that we can clean up the Cuyahoga, we can be assured that we are capable of cleaning up any river in the Nation.

In June of 1969, the river actually caught fire, causing almost $100,000 damage to two railroad bridges. A continuous and vigorous cleanup program could have prevented this shameful occurrence.

At this time, there is virtually no fish life in the lower Cuyahoga; in fact, hardly any biological life at all. Dissolved oxygen—the life-giving substance of all biological life—has been depleted.

Recreational uses of Lake Erie have also decreased because of the deterioration of lake quality caused by the river. Although these losses, both in terms of real loss and unrealized economic increase, are difficult to quantify, a decline has been evident. Sport fishing has been affected by the decrease in game fish abundance in the same way commercial fishing operations have suffered. In addition, swimming has been disallowed in several areas because of the hazardous levels of bacterial contamination.

In short, the rape of the Cuyahoga River has not only made it useless for any purpose other than a dumping place for sewage and industrial waste, but also has had a deleterious effect upon the ecology of one of the Great Lakes.

We can, however, take a significant step forward in the effort to cleanse the Cuyahoga with the passage of this bill. We can, by removing debris from the river and its banks, by dredging and scalping the river banks, and by embarking upon a bank stabilization program, prevent hundreds of tons of debris and suspended solids material from entering Lake Erie. Thus, we can make an important contribution to the improvement of what may well be the Nation's most critical pollution problem area.

RAY OSRIN

"SOMEDAY SON, ALL THIS WILL BE YOURS"

CLEVELAND *PLAIN DEALER* (AUGUST 16, 1970)

'SOMEDAY SON, ALL THIS WILL BE YOURS.'

ELEANOR PHINNEY

Homeowners Preservation League

LETTER TO THE OREGON ENVIRONMENTAL COUNCIL

JUNE 12, 1970

Oregon Environmental Council
1238 NW Glisan
Portland, OR 97204

Dear Sir:

Ecology has recently emerged from a long-ignored body of information, and our environment has become a national concern of major proportions. It would appear that we must take action now if we want to preserve any sort of livability. The damage to our land, air and water has been extensive in many cases, but it is not irreversible. Reclamation could be a trial of a thousand crimes.

But instead of sitting in judgment on the past abuses of our lovely valley, the members of our league are proposing a creative, vital effort to save our land, and design its use to be in harmony with nature. A model study for land use planning has been taken on as a classroom project at Oregon State University. It will use ecological determinants for the analysis of the land use planning for future zoning potential. Last Monday we saw the preliminary work that has been done by a group of professors and students and it is very encouraging. We are grateful for the assistance and cooperation of many agencies from all levels of government in our collection of pertinent data and studies vital to this study. It is our sincere hope that these efforts will ultimately save the intrinsic beauty of our valley and that the model study will serve as a guide to establish standards for Land Use by which we can achieve a rejuvenation of our environment.

Oregon Environmental Council Records, box 2, folder 36, Oregon Historical Society.

The following outline will give you some idea of the approach we are taking as we humbly submit the proposal of the Conference and program for your consideration and suggestions for improvement. We hope the potential will interest you and your associates and that you will join us as one of the co-sponsors of the finalized Conference program.

Proposal for Oregon's Environment Planning Conference
Fall, 1970

I. Evolution of Oregon Territory (Pacific Northwest): Presentation of visual aids (filmed if possible) and narrative of the evolution of Oregon's geologic development, the evolution of plant and animal life and explanation of their interdependence, and man's intrusion on the scene, followed by his inventions, leading to our present environment and problems.

II. Film of Model Study of the Lower Tualatin Valley—a total environmental analysis for Land Use Planning and Zoning by ecological determinants.

    A. Geological factors: hazards, flood plains, water resources, soils analysis

    B. Animal and plant life, and natural resources necessary to sustain life cycles

    C. Meteorology—air flow, inversion frequency, pollutants

    D. Analysis of existing impactors: Tualatin River, highways and roads, public utilities, sanitary facilities, historical sites, homes, schools, commercial establishments, industry, and agriculture

    E. Limiting factors for future development

    F. Future development potential: optimum population attainable using:

        1. Well water and irrigation from river;

        2. Imported water;

        3. Imported water, storm drains and sewers

    G. Economic and sociological benefits of planned growth (open end) versus urban sprawl

III. Open Forums and Symposiums at the Conference. Public participation by question-and-answer periods at conclusion of session.

    A. Technical assistance for planners interested in using ecological determinations in land use planning and zoning. Schedule by regions: Coastal, Interior Valley, Desert.

    B. Presentations of New Town developments and land use planning of other states and foreign countries to establish precedence and to interpret their successes and suggest improvements.

    1. England, Sweden, France, Puerto Rico (development of new towns)
    2. Wisconsin, Hawaii, Regional Plan for Lake Tahoe
  c. New legislation to improve and protect the environment resulting from concept of ecological land use determination; exploration of different methods of taxation.
  D. Public education presentations on conservation practices:
    1. Preservation of Natural Resources and Wildlife
    2. Agriculture, pesticides, fertilizers, waste disposal, crop rotation and contouring
    3. Forestry, public parks, litter, vandalism problems and costs
    4. Transportation, highways, roads, mass transportation and the automobile
    5. Solid waste disposal and recycling
    6. Management of private and public landscapes techniques

A Conference along the lines outlined above with the support of all factions of our citizenry—private, residential, industrial, commercial, and the public agencies and educational institutions—could be the start of a combined effort which is so badly needed to preserve and maintain the beauty of our land and natural resources. We respectfully request your consideration and cooperation toward this end.

Sincerely,

Homeowners Preservation League
Eleanor Phinney, Chairman
Environmental Conference
Rt. 1, Box 500
West Linn, Oregon 97068

PUBLIC SERVICE ANNOUNCEMENTS

APRIL 24–SEPTEMBER 21, 1970

10 SECOND PUBLIC SERVICE ANNOUNCEMENT

Don't let anyone kid you. Industry has the technical know-how to clean up the air in Pittsburgh. But will they use it? Don't hold your breath. FIGHT FOR IT. Write GASP, Box 2850, Pittsburgh, Pa. 15230.

30 SECOND PUBLIC SERVICE ANNOUNCEMENT

Have you heard about GASP—Your citizens Group Against Smog and Pollution? It's leading the fight to clean up the air in Allegheny County—before it's too late. GASP attends the variance hearings and has expert ecologists who meet every cop-out industrial polluters have to offer. They have a speakers' bureau and a staff of top flight advisors who volunteer their expertise against polluted air. GASP wants clean air for you and for your children. But will the polluters let you have it? Don't hold your breath. FIGHT FOR IT. Join GASP, write Box 2850, Pittsburgh 15230.

40 SECOND PUBLIC SERVICE ANNOUNCEMENT

The young and the old. Babies and children. Grandmothers and grandfathers. These are the unfortunates who are affected most by the killer gases in Pittsburgh's polluted air.

You are living in the sixth most air polluted city in the United States. Yes, Pittsburgh was cited by the federal government as having one of the nation's six biggest pollution problems.

Who is pouring the killer gas, sulfuric dioxide into your children's lungs? A leading environmental scientist says that essentially all of the sulfur in the air of Allegheny County comes from the steel mills and the power plants. Can

Michelle Madoff Collection, box 2, folder 22, at the University of Pittsburgh Archive.

you support life without clean air? Don't hold your breath. Join GASP, the Pittsburgh-based citizens' Group Against Smog and Pollution. Write P.O. Box 2850, Pittsburgh 15230.

60 SECOND PUBLIC SERVICE ANNOUNCEMENT

How many times have you heard the story that we cleaned up Pittsburgh years ago? Do you know that Pittsburgh air is far more dangerous to breathe now than it was when you couldn't see the sun at noon? The fly ash and other large dust particles that settle on your car, your window sills and your furniture is holding constant.

But the danger is the gas you do not see—the sulfur dioxide that our environmental scientists tell us is increasing at all Allegheny County locations. . . . Sulfur dioxide is a gas that pours out of the tall smoke stacks of power plants and steel mills. Sulfur dioxide turns into sulfuric acid when it hits the open air and it eats away at the bricks of your home and at the metal on your car. It attacks your lungs, your eyes and your skin.

The present high level of sulfur dioxide in certain parts of Allegheny County aggravates lung disease, emphysema and bronchitis and doctors are now linking air pollution with pneumonia and lung cancer.

The power plants and steel mills can do something. They have the technology. You can do something, too. Write GASP, Box 2850, Pittsburgh 15230.

20 SECOND PUBLIC SERVICE ANNOUNCEMENT

If you think the auto is Allegheny County's major source of air pollution, you're wrong. The auto is the villain in Los Angeles but it's steel mills, utility companies and industrial plants here. Demand enforcement of Allegheny County's anti-pollution laws. Write GASP, Box 2850, Pittsburgh 15230.

PITTSBURGH—Mothers are all alike. They spend most of the day washing clothes, washing dishes, washing diapers, dusting and cleaning and scrubbing. A clean house means a healthy family.

But what about the air? Is someone else out there scrubbing and cleaning the air? Don't hold your breath! FIGHT FOR IT. Attend the public meeting of GASP, Pittsburgh's group against smog and pollution, at 8 p.m. on Thursday, October 22, in the Graduate School of Public Health auditorium, Oakland.

CLEAN AIR ACT AMENDMENTS

DECEMBER 31, 1970

AN ACT

To amend the Clean Air Act to provide for a more effective program to improve the quality of the Nation's air.

Air Quality Control Regions

SEC. 107.

(a) Each State shall have the primary responsibility for assuring air quality within the entire geographic area comprising such State by submitting an implementation plan for such State which will specify the manner in which national primary and secondary ambient air quality standards will be achieved and maintained within each air quality control region in such State.

(b) For purposes of developing and carrying out implementation plans under section 110—

    (1) an air quality control region designated under this section before the date of enactment of the Clean Air Amendments of 1970, or a region designated, as such date under subsection (c), shall be an air quality control region; and

    (2) the portion of such State which is not part of any such designated region shall be an air quality control region, but such portion may be subdivided by the State into two or more air quality control regions with the approval of the Administrator.

(c) The Administrator shall, within 90 days after the date of enactment of the Clean Air Amendments of 1970, after consultation with appropriate State and local authorities, designate as an air quality control region any inter-

---

Excerpts from Public Law 91–604.

state area or major intrastate area which he deems necessary or appropriate for the attainment and maintenance of ambient air quality standards. The Administrator shall immediately notify the Governors of the affected States of any designation made under this subsection.

Air Quality Criteria and Control Techniques
SEC. 108. (a) (1) For the purpose of establishing national primary and secondary ambient air quality standards, the Administrator shall within 30 days after the date of enactment of the Clean Air Amendments of 1970 publish, and shall from time to time thereafter revise, a list which includes each air pollutant—

> (A) which in his judgment has an adverse effect on public health or welfare;
> (B) the presence of which in the ambient air results from numerous or diverse mobile or stationary sources; and
> (C) for which air quality criteria had not been issued before the date of enactment of the Clean Air Amendments of 1970, but for which he plans to issue air quality criteria under this section.

(2) The Administrator shall issue air quality criteria for an air pollutant within 12 months after he has included such pollutant in a list under paragraph (1). Air quality criteria for an air pollutant shall accurately reflect the latest scientific knowledge useful in indicating the kind and extent of all identifiable effects on public health or welfare which may be expected from the presence of such pollutant in the ambient air, in varying quantities. The criteria for an air pollutant, to the extent practicable, shall include information on—

> (A) those variable factors (including atmospheric conditions) which of themselves or in combination with other factors may alter the effects on public health or welfare of such air pollutant;
> (B) the types of air pollutants which, when present in the atmosphere, may interact with such pollutant to produce an adverse effect on public health or welfare; and
> (C) any known or anticipated adverse effects on welfare.

(b) (1) Simultaneously with the issuance of criteria under subsection (a), the Administrator shall, after consultation with appropriate advisory committees and Federal departments and agencies, issue to the States and appropriate air pollution control agencies information on air pollution control techniques, which information shall include data relating to the technology and costs of emission control. Such information shall include such data as are available on available technology and alternative methods of prevention and control of air pollution. Such information shall also include data on alternative fuels, processes, and operating methods which will result in elimination or significant reduction of emissions. . . .

Motor Vehicle Emission Standards
SEC. 202. (a) Except as otherwise provided in subsection (b)—

(1) The Administrator shall by regulation prescribe (and from time to time revise) in accordance with the provisions of this section, standards applicable to the emission of any air pollutant from any class or classes of new motor vehicles or new motor vehicle engineers, which in his judgment causes or contributes to, or is likely to cause or to contribute to, air pollution which endangers the public health or welfare. Such standards shall be applicable to such vehicles and engines for their useful life (as determined under subsection [d]), whether such vehicles and engines are designed as complete systems or incorporated devices to prevent or control such pollution.

(2) Any regulation prescribed under this subsection (and any revision thereof) shall take effect after such period as the Administrator finds necessary to permit the development and application of the requisite technology, giving appropriate consideration to the cost of compliance within such period.

(b) (1) (A) The regulations under subsection (a) applicable to emissions of carbon monoxide and hydrocarbons from light duty vehicles and engines manufactured during or after model year 1974 shall contain standards which require a reduction of at least 90 per centum from emissions of carbon monoxide and hydrocarbons allowable under the standards under this section applicable to light duty vehicles and engines manufactured in model year 1970.

(B) The regulations under subsection (a) applicable to emissions of oxides of nitrogen from light duty vehicles and engines manufactured during or after model year 1976 shall contain standards which require a reduction of at least 90 per centum from the average of emissions of oxides of nitrogen actually measured from light duty vehicles manufactured during model year 1971 which are not subject to any Federal or State emission standard for oxides of nitrogen. Such average of emissions shall be determined by the Administrator on the basis of measurements made by him. . . .

Motor Vehicle and Motor Vehicle Engine Compliance Testing and Certification
SEC. 206. (a) (1) The Administrator shall test, or require to be tested in such manner as he deems appropriate, any new motor vehicle or new motor vehicle engine submitted by a manufacturer to determine whether such a vehicle or engine conforms with the regulations prescribed under section 202 of this Act. If such vehicle or engine conforms to such regulations, the Administrator shall issue a certificate of conformity upon such terms, and for such period (not in excess of one year), as he may prescribe.

(2) The Administrator shall test any emission control system incorporated in a motor vehicle or motor vehicle engine submitted to him by any person, in order to determine whether such system enables such vehicle or engine to conform to the standards required to be prescribed under section 202(b) of this Act. If the Administrator finds on the basis of such tests that such vehicle or engine conforms to such standards, the Administrator shall issue a verification of compliance with emission standards for such system when incorporated in vehicles of a class of which the tested vehicle is representative. He shall inform manufacturers and the National Academy of Sciences, and make available to the public, the results of such tests. Tests under this paragraph shall be conducted under such terms and conditions (including requirements for preliminary testing by qualified independent laboratories) as the Administrator may prescribe by regulations.

PART 4

IS CATASTROPHE COMING?

THE DOCUMENTS COLLECTED IN THIS SECTION, LARGELY FROM 1971 and 1972, capture widely varying thoughts on the severity of the nation's environmental problems. Although some Americans clearly felt a sense of crisis, others considered much of the movement's rhetoric alarmist. Not surprisingly, Americans also held a wide range of opinions on how much action would be necessary to solve the problems that did exist. How much regulation would be necessary? How many resources should society invest in environmental protection? Who should pay for environmental improvements? Should Americans trust government, especially the federal government, to solve environmental problems? As these documents reveal, criticism of the movement and its successes came from both the political left and from the growing conservative movement, even as calls for better environmental policy continued.

The first two documents concern clear-cut forestry in the Bitterroot Mountains of Montana. In 1970 the University of Montana published a report, excerpted in this section, on national forest policy. The report concluded that the Forest Service was having difficulty changing its culture and policies to match public expectations, which themselves were rapidly evolving because of the environmental movement and the spread of ecological understanding. In November of 1971 the *New York Times* ran an article about the national forests, in which the paper's environmental writer, Gladwin Hill, declared that the nation's greatest reservoir of material and recreational resources were "in trouble," noting that bulldozers were "boring into some of the nation's last remaining pristine wilderness." Accompanying the article, which began on page one, was a photograph by Dale A. Burk, taken in the Bitterroot Forest.

The photograph, included here, featured forester G. M. Brandborg and Wyoming senator Gale W. McGee, but it mostly revealed the nature of modern clear-cut forestry and the terracing technique that theoretically allowed more efficient replanting. These were the very practices criticized in the University of Montana report published the previous year. The Burk photograph reminds us of the importance of images in shaping public opinion on the environment.

The third document comes from Ronald Reagan, at the time governor of California. In a talk before the American Petroleum Institute, Reagan described his increasing concern about environmental alarmism and the growth of government bureaucracy. Although Reagan did not dismiss environmental problems, he suggested that industry could regulate itself and in fact had already made great strides in protecting the environment. Nine years after this speech, Reagan was elected president, riding the rising tide of New Right politics and indeed shaping the movement with his antigovernment rhetoric.

Another skeptic, Dr. Joseph T. Ling, an executive at the 3M Company and a representative of the National Association of Manufacturers, argued before the House Committee on Public Works that industry cannot reasonably remove all pollutants from its effluent, claiming that to do so would do more harm than good. In his testimony before Congress as it considered the bill that ultimately became the Clean Water Act (1972), Ling asserted that regulators needed to balance the benefits of eliminating pollution with the costs of doing so. This cost-benefit rhetoric would become increasingly important throughout the 1970s and 1980s. His testimony also reminds us that effective pollution regulation required scientific study and the sharing of data and conclusions with policy makers and the public.

The fourth document comes from Kentucky's coal country, where residents have been engaged in a long struggle to stop strip mining and, more recently, mountain-top removal. The prose included here comes from a pamphlet entitled "We Will Stop the Bulldozers," which describes the actions taken by a number of women in Eastern Kentucky. It reveals how much environmental activists learned from the labor and civil rights movements, and it suggests how radical average people could sound when protecting their homes and communities from "destruction for profit's sake."

Supreme Court Justice William O. Douglas was an outspoken advocate for the environment, and his position on the court gave him the opportunity to influence the law and the national discourse. One of his most important contributions to the latter came in his dissent in the Mineral King case of

1972 (*Sierra Club v. Morton*). The court ruled that the Sierra Club could not sue to protect Mineral King from development, which the National Forest Service favored, but that individual members could sue. Douglas went much further in his dissent, claiming that Mineral King itself should have standing to sue, or at least people who have a connection to it, or any place, should have standing. In the process Douglas wrote one of the most eloquent and radical defenses of nature to date.

In 1972, John Maddox was the editor of *Nature*, the prominent science journal, and the author of the popular book *The Doomsday Syndrome* in which he argued against the "prophets of doom." Claiming that the world's food problem would be surpluses rather than shortages, Maddox takes on Paul Ehrlich directly in the excerpt included in this section. His book more broadly attacked negative rhetoric surrounding the environmental movement; Maddox placed his faith in human innovation and technological development.

The final document in this section comes from the United Nations Conference on the Human Environment, held in Stockholm, Sweden, in June 1972. Representatives from more than one hundred countries attended this first ever United Nations conference on the global environment. The conference led to a declaration, excerpted in this section, that listed common principles and beliefs concerning human interaction with the environment in an increasingly crowded world. The document is based in the language of compromise and moderation, but it clearly reveals a broad understanding of the need for governments and individuals to "exert common efforts for the preservation and improvement of the human environment."

REPORT ON THE BITTERROOT NATIONAL FOREST

1970

STATEMENT OF FINDINGS

1. Multiple use management, in fact, does not exist as the governing principle on the Bitterroot National Forest [BNF].
2. Quality timber management and harvest practices are missing. Consideration of recreation, watershed, wildlife and grazing appear as afterthoughts.
3. The management sequence of clear-cutting-terracing-planting cannot be justified as an investment for producing timber on the BNF. We doubt that the Bitterroot National Forest can continue to produce timber at the present harvest level.
4. Clear-cutting and planting is an expensive operation. Its use should bear some relationship to the capability of the site to return the cost invested.
5. The practice of terracing on the BNF should be stopped. Existing terraced areas should be dedicated for research.
6. A clear distinction must be made between timber *management* and timber *mining*. Timber *management*, i.e., continuous production of timber crops, is rational only on highly productive sites, where an appropriate rate of return on invested capital can be expected. All other timber cutting activities must be considered as timber mining.
7. Where timber mining, i.e., removing residual old growth timber from sites uneconomical to manage, is to be practiced, all other on-site values must be retained. Hydrologic, habitat, and aesthetic values must be preserved by single-tree selection cutting, a minimum disturbance of all residual vegetation, and the use of a minimum standard, one-time, temporary road.

The Bolle Collection, box 67, folder 7, Archives and Special Collections, at the University of Montana, excerpts from pages 1–4.

8. The research basis for management of the BNF is too weak to support the management practices used on the forest.

9. Unless the job of total quality management is recognized by the agency leadership, the necessary financing for the complete task will not be aggressively sought.

10. Manpower and budget limitations of public resource agencies do not at present allow for essential staffing and for integrated multiple-use planning.

11. Present manpower ceilings prevent adequate staffing on the BNF. Adequate staffing requires people professionally trained and qualified through experience.

12. The *quantitative* shortage of staff specialists will never be resolved unless the *qualitative* issue with respect to such specialists is first resolved.

13. We find the bureaucratic line structure as it operates archaic, undesirable and subject to change. The manager on the ground should be much nearer the top of the career ladder.

14. The Forest Service as an effective and efficient bureaucracy needs to be reconstructed so that substantial, responsible, local public participation in the processes of policy-formation and decision-making can *naturally* take place.

15. It appears inconceivable and incongruous to us that at this time, with the great emphasis upon a broad multiple-use approach to our natural resources—especially those remaining in public ownership—that any representative group or institution in our society would advocate a dominant-use philosophy with respect to our natural resources. Yet it is our judgment that this is precisely what is occurring through the federal appropriation process, via executive order and in the Public Land Law Review Commission's Report. It would appear to us that at this time any approach to public land management which would de-emphasize a broad multiple-use philosophy, a broad environmental approach, a broad open-access approach, or which would reduce the production of our public land resources in the long run is completely out of step with the interests and desires of the American people. What is needed is a fully funded program of action for quality management of all of our public lands.

The problem arises from public dissatisfaction with the Bitterroot National Forest's overriding concern for sawtimber production. It is compounded by an apparent insensitivity to the related forest uses and to the local public's interest in environmental values.

In a federal agency which measures success primarily by the quantity of timber produced weekly, monthly, and annually, the staff of the Bitterroot National Forest finds itself unable to change its course, to give anything but token recognition to related values, or to involve most of the local public in any way but as antagonists.

The heavy timber orientation is built in by legislative action and control, by executive direction and by budgetary restriction. It is further reinforced by the agency's own hiring and promotion policies and it is rationalized in the doctrines of its professional expertise.

This rigid system developed during the expanded effort to meet national housing needs during the post-war boom. It continues to exist in the face of a considerable change in our value system—a rising public concern with environmental quality. While the national demand for timber has abated considerably, the major emphasis on timber production continues.

The post-war production boom may have justified the single-minded emphasis on timber production. But the continued emphasis largely ignores the economics of regeneration; it ignores related forest values; it ignores local social concerns; and it is simply out of step with changes in our society since the post-war years. The needs of the post-war boom were met at considerable social as well as economic cost. While the rate and methods of cutting and regeneration can be defended on a purely technical basis, they are difficult to defend on either environmental or long-run economic grounds.

Many local people regard the timber production emphasis as an alien orientation, exploiting the local resource for non-local benefit. It is difficult for them to distinguish what they see from the older forest exploitation which we deplored in other regions. They feel left out of any policy formation or decision-making and so resort to protest as the only available means of being heard.

Many of the employees of the Forest Service are aware of the problem and are dissatisfied with the position of the agency. They recognize the agency is in trouble, but they find it impossible to change, or, at least, to change fast enough.

Multiple-use is stated as the guiding principle of the Forest Service. Given wide lip-service, it cannot be said to be operational on the Bitterroot National Forest at this time.

A change in funding to increase considerably the activities in non-timber uses would help, but could not be effective until legislative and executive emphasis is changed.

But even with this modification the internal bureaucracy of the agency and the lack of public involvement in decision-making make real change unlikely.

As long as short-run emphasis on timber production overrides long-run (and short-run) concern for related uses and local environmental quality, real change is impossible and the outlook is for continued conflict and discontent.

DALE A. BURK

PHOTOGRAPH OF THE BITTERROOT FOREST, MONTANA

NEW YORK TIMES (NOVEMBER 19, 1971)

Photo provided by and used with the permission of Dale A. Burk.

RONALD REAGAN

REMARKS BEFORE THE AMERICAN PETROLEUM INSTITUTE

SAN FRANCISCO, NOVEMBER 16, 1971

It seems only yesterday we were hearing a great hue and cry about a scheduled underground nuclear test on a remote Alaskan island.

There were dire predictions that great earthquakes and tidal waves would create havoc as far away as Hawaii and Japan. One could not help but think of those groups who on occasion take to a hilltop to await the end of the world. Only these latter-day doomcriers are knowledgeable and seemingly responsible citizens who offered their dire predictions of an almost mortal blow against the environment without one shred of scientific evidence to prove their claims. The test went off on schedule without earthquake or tidal wave. Officials monitoring the scene have yet to detect any radiation in the atmosphere.

But there has been a strange silence from those who objected most vigorously and vociferously. I have yet to read or hear of any of them holding a press conference to announce that they were wrong, that it is possible for America—without causing environmental ill-effects—to test an anti-nuclear defense system that may someday prove crucial to the nation's survival.

If we let our memory go back a little farther to a place called Bikini—when that was an island in the Pacific, not a mini size bathing suit—we recall some genuinely alarmed citizens who thought that test would blow a hole in the bottom of the ocean and let all the water drain out.

The recent Amchitka controversy is another example of something that might be called the Doomsday syndrome so prevalent in our country in recent years. There is of course a new awareness of nature and our responsibility to preserve the beauty and the wonder of this spaceship called earth. I know few, if any, who don't feel this way.

Governor's Papers: Press Unit, box P18, folder "Speeches—Governor Ronald Reagan, 1971," Ronald Reagan Library, excerpts from pages 1–5 and 10.

Protecting the environment now receives a high priority in almost every industrial and individual activity, yet the Doomsday crowd is not satisfied. Their exaggerations hurt the cause of the sincere and dedicated conservationists who have done so much to alert us to the need for environmental safeguards.

Their pervasive pessimism is anti-technology, anti-industry and includes opposition to the defense program we must have to maintain the very freedom that allows them to speak their minds and stage their demonstrations. From all this has come a downgrading and even a reviling of the most prosperous and advanced society in the world.

A free enterprise system that has given America the highest standard of living in the world is portrayed as a conspiracy against the poor.

A technology that allows the average American to live better, longer and with more conveniences than the wealthiest monarch could afford 50 years ago is denounced at worst as a tool of the so-called "military-industrial" complex, at best as an evidence of our crass materialism. Energy sources that fuel our homes, our transportation systems, the industries employing our people, are attacked as massive threats to the environment.

Our system of government is accused of repression, of denying either economic or social equality to minorities and of not caring about injustice or the poor and hungry.

We have always had prophets of doom and gloom with us. But their ranks have proliferated.

And because of television and other technological advances which some of them regard as socially menacing, they are able to spread their pessimistic view of things to every corner of the globe.

We seem to live in an age of simplistic overstatement and false propaganda.

We used to have problems. Today, we have crises.

Worry about over-population is twisted and projected into a threat of imminent mass starvation.

Education, the effort to end discrimination, our health needs, almost every valid concern of a forward-looking and humanitarian society become causes around which the Doomsday crowd rallies to spin their tale of calamity.

Somehow, they always seem to ignore the very real progress we have made in meeting the needs of our people.

Your industry has been plagued by the Doomsday syndrome as much or more than most. Yet, those of you who produce the nation's oil and petroleum products share the determination of our people to end air and water pollution and to stop destructive environmental practices.

Our own state has led the nation, indeed the world, in efforts to protect the environment against everything from smog to offshore oil spills.

We have enacted and are enforcing the nation's strictest water and air pollution controls. And, we are convinced that industrial progress can be made compatible with the necessary efforts to protect the environment. Petroleum is California's Number One mineral commodity. Its annual value of $1.2 billion exceeds the value of all other mineral production combined. More than 600,000 of California's 20 million people derive their livelihood directly from the petroleum industry.

Oil and petroleum products fuel the cars, trucks, tractors, buses and airplanes we use to ride to work, and produce our food. Oil products provide part of the power for the industries which give employment to our people and for the hospitals which heal them when they are sick.

It has been said that a modern economy literally runs on oil and California is no exception. Yet I am told that paying compliments to your industry is not the smartest thing politically a fellow can do in today's climate. As a matter of fact, you are almost as picked on as actors used to be.

Well, take heart—if worst comes to worst, you can always try politics. . . .

To provide and maintain the kind of environmental safeguards we must have, your industry we know will display the kind of constructive attitude it has demonstrated during environmental problems of the past. Too often, your costly and commendable efforts have been ignored by critics eager to cast the oil industry as the Number One environmental villain.

Following the off-shore oil blow-outs on federal leases off Southern California two years ago, the firms involved did not wait for government order. Without hesitation, they started to clean up the beaches . . . at an estimated cost of $10 million.

When our State Department of Fish and Game noted a loss of wildlife in oil sumps in the Southern San Joaquin Valley, the oil industry without any government coercion started a massive cleanup campaign. More than 800 sumps have been filled or covered at a cost of about $750,000.

When the U.S. Navy accidentally spilled 5,000 barrels of oil in a refueling operation off Southern California this year, the industry sent advisors to assist in the clean-up.

Your response to the public demand for environmental safeguards has not been limited to a reaction after the problem occurs.

You have invested millions of dollars for new and more effective equipment to control air and water pollution and to make refinery and other operations compatible with the natural environment.

Standard Oil of California, headed by Chairman Otto Miller, has removed more than 3,000 advertising billboards to help enhance and preserve the scenic beauty of our rural landscapes in California. Other firms have taken similar steps. Since the Doomsday myth-makers rarely mention this, I thought I would.

But you will be hearing from the experts about the problems affecting your industry. I would like to spend a few moments examining a few widely accepted Doomsday myths to see how they stand up to a few facts.

Maybe it is hard for us to recall some of our childhood fears and how very real they were in the dark of night. I receive a great many letters from children—sometimes from a whole class telling me of their belief that unless someone does something, they will die before they can grow up because there will be no air—or the water will be poisoned. They ask if it is true that all the trees will be gone in a few years. One whole class was convinced we would be making plastic trees to replace our once great forests. The Doomsday myth-makers produce a peculiar smog of their own.

Population control is one of their popular causes. Zero population growth is the rallying cry. The specter of mass starvation, of people standing elbow-to-elbow . . . is raised as the frightening prospect if we do not take drastic steps to curb the birth rate. Some of the steps proposed involve a kind of regimentation Americans have always found unacceptable.

Never mind if the plain, unvarnished truth about our population growth makes their rhetoric sound a little melodramatic and downright silly. Despite all the furor, the United States is not producing a bumper baby crop every year.

In fact, after reaching a peak of 3.8 children per average family in 1957, the birth rate in America has been declining steadily ever since. It is now estimated at 2.3 children per family. . . .

Environmentalists delight in quoting Thoreau to bolster their case. I hope they won't mind my using him for the same reason. He said, "Yet this government of itself never furthered any enterprise except when it got out of its way. The character inherent in the American people has done all that has been accomplished; and it would have done somewhat more if government had not sometimes got in the way."

Government and business working together—each in its proper place—makes for an irresistible force. One half of the economic activity of the entire human race has been conducted under American auspices. No other system can even begin to match our abundance. . . .

This is the most dynamic, humane, forward-looking society in the world. We do care about the oppressed, the disadvantaged, the minorities. Freedom and individual dignity are as important to us as the technology that made them possible.

Whatever the Doomsday myth-makers say, this is the brightest hope of men who seek a brighter tomorrow.

The next time you are told how much better government can make things if only government had a little more power and resources, refer them to that great nation which has practiced total government control without interference for more than half a century.

We can, if we are willing to expend the effort, match the economic achievements of the Soviet Union. It would mean moving 60 million Americans back to the farm, abandoning 60 percent of the steel industry and 2/3 of the oil industry. We would junk 90 percent of our cars and our telephones, rip up 2/3 of the railroad tracks and tear down 70 percent of our houses.

There would be only one thing left to do then to match their government-run paradise; give up 100 percent of our freedom.

(NOTE: Since Governor Reagan speaks from notes, there may be changes in, or additions to, the above quotes. However, the governor will stand by the above quotes.)

DR. JOSEPH T. LING

TESTIMONY REGARDING THE WATER POLLUTION CONTROL ACT

Mr. Chairman, and members of the committee, as a technical man, I have been working in the water pollution field for 25 years, as an educator, researcher, and engineer.

When I first heard about the zero discharge, the first thing that came to my mind was not money; it was what is the environmental impact?

I mean the environmental impact because we are doing something to remove something from the system, because we are using a lot of energy and material to remove the last trace of pollutants from the water, because we cannot destroy material and we can only change the material from one form to another; and therefore what I was worried about, when we go to the zero discharge, will we really take more pollution out of the environment than we put in?

If we put more pollution into the environment no matter if it is here or elsewhere, no matter if it is water pollution or air pollution, I call this a negative environmental impact. If we can take out of the environment more pollutants than we put in, I call it a positive environmental impact.

My major concern is, what is the environmental impact of the zero discharge? We did some figuring at one of our 3M company manufacturing plants. I must confess to you, gentlemen, that I had no intention of releasing this information to the public until I learned the answer.

After I learned the answer, I thought the answer was very important, and, as a professional person, I figured it was my obligation to make this information available to this distinguished committee for your consideration.

First, what is the zero discharge? From my definition, as a technical man, we can have no zero. The discharge we are using for these calculations is based

House Committee on Public Works, *Water Pollution Control Legislation—1971*: *Hearings on H.R. 11896*, 92nd Cong., 1st sess., 1971, pages 635–37.

on the U.S. Drinking Water Standards, published by the U.S. Public Health Service. For this particular question, I tried to purify the water from this particular plant to meet the same quality as the drinking water standards.

To do this, I had to exercise a number of unit operations, such as distillation, neutralization, stripping, condensation, secondary activated sludge treatment, and so on, and so forth.

In order to do this, we would have to spend $25 million for capital investment and $3.5 million at least for maintenance and operation per year.

But that is not the point. The point is that, while we are using this money, we have to buy necessary equipment, concrete and steel, to build this facility. In addition to that, we have to purchase 9,000 tons of chemicals to make this operation run. This includes sulfuric acid, and caustic carbon, and so on, and so forth.

In addition to these 9,000 tons of chemicals, we have to purchase approximately 1,500 kilowatts of electricity.

Also, we have to use 19,000 tons of coal to produce the steam for this particular operation. We figured we probably would remove about 4,000 tons of pollutants from the water of this plant.

In order to do this operation, we would produce 9,000 tons of chemical sludge and about 1,200 tons of fly ash from the boiler, 1,000 tons of sulfur dioxide, and 200 tons of nitrogen oxide in terms of air pollution.

This is only about the 3M Co. How about the 9,000 tons of chemicals purchased from someone else in the country? In order, then, for him to produce 9,000 tons of chemicals to use for this particular process, according to the *Encyclopedia of Chemical Technology*, 1967 edition, I made a very quick calculation.

He would need 15,000 tons of natural resources to produce this 9,000 tons of chemicals, and he would also need additional power to do that job.

Meanwhile, he would produce 6,500 tons of sludge. That is, solid wastes.

How about the 1,500 kilowatts of electricity? According to the data published by the U.S. Public Health Service, he would have to use 6,000 tons of coal to produce these 1,500 kilowatts. Meanwhile, he would have to emit 350 tons per year of sulfur dioxide, 60 tons of fly ash, and 60 tons of nitrogen oxide, plus 100 million BTUs per year of waste heat to be disposed of somehow, either to the air or to the river.

How about the steel supply that we bought for the steel tank and the concrete tank? I would have to go to the same calculation for the steel manufacturing and the cement operations, which I did not do.

Just use of these major cycles that I draw to the conclusion. In order to remove approximately 4,000 tons of pollutants from this particular plant, we would have to use more than 40,000 tons of natural resources.

We would produce approximately 19,000 tons of waste material in terms of solid wastes or air pollution. That is four times as much as we removed from our plant.

In addition to that $25 million capital plus $3.5 million for annual operation, I think the most important thing is the present effluent from our so-called equivalent secondary treatment plant is meeting the Federal and State water quality standards.

My conclusion is that the zero discharge based on this particular operation would produce a negative environmental impact. If you looked into 3M's effluent pipe, yes, you would get a clear effluent. But you go up a little bit higher to look at the overall environment for the country and you would find that we created a lot more pollution than we have removed from this plant.

There must be an optimum environmental impact. This could be varied from 85 to 90 to 95 percent. My calculations indicated to me definitely that it is not "zero."

I think, "the cleaner the better," that is true. Only when we look into the "cleaner"—it is not "cleaner" from the pipe but is it "cleaner" for the overall environment of the country? So I feel that "zero discharge" is economically unwise and environmentally undesirable.

I do not want to imply this figure, or other numbers, as representing the total impact on the overall environment. I did not complete my calculations. I just wanted to bring this type of environmental impact to your consideration.

I do hope the committee will be able to dig into this type of information much more deeply and broadly before you make a decision.

Thank you very much.

WE WILL STOP THE BULLDOZERS

JANUARY 20, 1972

WE WILL STOP THE BULLDOZERS

At 6:00 this morning a group of women who are your neighbors stopped the Sigmon Brothers stripping operation at Elijah Fork on Trace Branch of Ball.

East Kentucky women of the Appalachian Group to Save the Land and People and of other groups are standing in front of the bulldozers. This is a non-violent demonstration. We feel that if our men would take the same action, the coal operators would unleash violence against them. But our men and children will be there to support us.

We are taking this action against the strip mining industry because we feel we have been given the runaround by our county, state, and federal officials. This week our elected officials are considering strip mining bills in Frankfort. We will continue to stop major strip mining operations to show Governor Ford and our Kentucky legislature that the people of Appalachia demand a total ban on strip mining. We also demand that the Knott County officials who passed a ban on strip mining one and a half years ago collect the hundreds of thousands of dollars in fines owed by the strippers for their illegal mining. That money belongs to the people whose land has been stripped in flagrant violation of the Knott County ordinance.

WE WILL NOT GIVE UP. WE NEED YOUR HELP AND SUPPORT!

All citizens of Knott County come and join us!

Bring food and warm clothing.

—The Women of Eastern Kentucky

Excerpts from pamphlet, Keith Dix, Strip Mining Papers, 1969–1973, West Virginia and Regional History Collection at the West Virginia University Libraries.

The above statement was prepared by the women who occupied the strip mine. It was distributed in mail boxes in Perry and Knott Counties to encourage more women to join the protest the next day, and to let people know the protest was happening.

WHICH SIDE ARE YOU ON?

We women who occupied the Ken Mack Coal Company's strip mine site sang a lot of songs during those 15 hours we were on the mountain. We sang songs to keep our spirits up and to keep up our strength. We were determined to be heard. One of the songs we sang was, "We Shall Not be Moved." That song meant a lot to us because we know that coal operators will do anything to defend their "right" to strip and destroy our land, and they will do anything to keep us from stopping them.

The coal operators stole the land from our grandparents many years ago. They paid only pennies an acre for the coal and land which has made millions of dollars of profit for them. Now they tear the land to ruin and destroy our homes and farms, while they pay gun thugs to keep us away. The operators pit us against each other—sister against sister, brother against brother, family against family, and neighbor against neighbor. They tell us that welfare recipients are no count, but they mine coal with as few workers as they can. They have a monopoly on the coal industry and have forced deep mines out of business, leaving many of our men unemployed. They tell their workers that people who fight to protect the land are keeping bread off the workers' tables, but they pay the few men they hire only a tiny bit of the millions they make. (In 1968 in Knott County, Ky., surface mining brought in $3 million, but employed only 100 men.) They spread lies among us to make us hate each other. They know the only way to keep their profits is to keep us divided.

But we shall not be moved. We will fight until all our people understand who the real enemy is. We will stand together against those who make millions by exploiting the coal miners and the poor. We will not tolerate destruction for profit's sake.

We sang another song up on that mountain, "Which Side Are You On?" This song too has a lot of meaning now. Ask this question of yourself as you read this pamphlet and think about what happened to us on January 20. The operators were determined to get rid of us, even to the point of unleashing violence. That violence sent four of our men to the emergency room of the hospital and caused extensive damage to two of our cars.

We are really in debt now. You can also take a stand by sending something, even if just a few pennies, to the Council of the Southern Mountains* CPO 2307, Berea, Ky. 40403. The Council is raising emergency money to help us pay these unexpected costs. Please help if you can.

We urge you to decide, "Which Side Are You On?"

* *tax deductible*

WILLIAM O. DOUGLAS

DISSENT, *SIERRA CLUB* V. *MORTON*

APRIL 19, 1972

I share the views of my Brother [Harry] BLACKMUN and would reverse the judgment below.

The critical question of "standing" would be simplified and also put neatly in focus if we fashioned a federal rule that allowed environmental issues to be litigated before federal agencies or federal courts in the name of the inanimate object about to be despoiled, defaced, or invaded by roads and bulldozers and where injury is the subject of public outrage. Contemporary public concern for protecting nature's ecological equilibrium should lead to the conferral of standing upon environmental objects to sue for their own preservation. See [Christopher D.] Stone, "Should Trees Have Standing?—Toward Legal Rights for Natural Objects," 45 *S. Cal. L. Rev.* 450 (1972). This suit would therefore be more properly labeled as *Mineral King* v. *Morton*.

Inanimate objects are sometimes parties in litigation. A ship has a legal personality, a fiction found useful for maritime purposes. The corporation sole—a creature of ecclesiastical law—is an acceptable adversary and large fortunes ride on its cases. The ordinary corporation is a "person" for purposes of the adjudicatory processes, whether it represents proprietary, spiritual, aesthetic, or charitable causes.

So it should be as respects valleys, alpine meadows, rivers, lakes, estuaries, beaches, ridges, groves of trees, swampland, or even air that feels the destructive pressures of modern technology and modern life. The river, for example, is the living symbol of all the life it sustains or nourishes—fish, aquatic insects, water ouzels, otter, fisher, deer, elk, bear, and all other animals, including man, who are dependent on it or who enjoy it for its sight, its sound, or its life. The

Supreme Court of the United States, 405 U.S. 727; 92 S. Ct. 1361; 1972 U.S. LEXIS 118. Footnotes have been removed.

river as plaintiff speaks for the ecological unit of life that is part of it. Those people who have a meaningful relation to that body of water—whether it be a fisherman, a canoeist, a zoologist, or a logger—must be able to speak for the values which the river represents and which are threatened with destruction.

I do not know Mineral King. I have never seen it nor traveled it, though I have seen articles describing its proposed "development" notably Hano, "Protectionists vs. Recreationists—The Battle of Mineral King," *N.Y. Times Mag.*, Aug. 17, 1969, p. 25; and Browning, "Mickey Mouse in the Mountains," *Harper's*, March 1972, p. 65. The Sierra Club in its complaint alleges that "one of the principal purposes of the Sierra Club is to protect and conserve the national resources of the Sierra Nevada Mountains." The District Court held that this uncontested allegation made the Sierra Club "sufficiently aggrieved" to have "standing" to sue on behalf of Mineral King.

Mineral King is doubtless like other wonders of the Sierra Nevada such as Tuolumne Meadows and the John Muir Trail. Those who hike it, fish it, hunt it, camp in it, frequent it, or visit it merely to sit in solitude and wonderment are legitimate spokesmen for it, whether they may be few or many. Those who have that intimate relation with the inanimate object about to be injured, polluted, or otherwise despoiled are its legitimate spokesmen.

The Solicitor General, whose views on this subject are in the Appendix to this opinion, takes a wholly different approach. He considers the problem in terms of "government by the Judiciary." With all respect, the problem is to make certain that the inanimate objects, which are the very core of America's beauty, have spokesmen before they are destroyed. It is, of course, true that most of them are under the control of a federal or state agency. The standards given those agencies are usually expressed in terms of the "public interest." Yet "public interest" has so many differing shades of meaning as to be quite meaningless on the environmental front. Congress accordingly has adopted ecological standards in the National Environmental Policy Act of 1969, Pub. L. 91–190, 83 Stat. 852, 42 U. S. C. § 4321 *et seq.*, and guidelines for agency action have been provided by the Council on Environmental Quality of which Russell E. Train is Chairman. See 36 Fed. Reg. 7724.

Yet the pressures on agencies for favorable action one way or the other are enormous. The suggestion that Congress can stop action which is undesirable is true in theory; yet even Congress is too remote to give meaningful direction and its machinery is too ponderous to use very often. The federal agencies of which I speak are not venal or corrupt. But they are notoriously under the control of powerful interests who manipulate them through advisory committees, or friendly working relations, or who have that natural affinity with the agency which in time develops between the regulator and the regulated. As early as 1894, Attorney General Olney predicted that regulatory agencies might become "industry-minded," as illustrated by his forecast concerning the Interstate Commerce Commission:

"The Commission . . . is, or can be made, of great use to the railroads. It satisfies the popular clamor for a government supervision of railroads, at the same time that that supervision is almost entirely nominal. Further, the older such a commission gets to be, the more inclined it will be found to take the business and railroad view of things." M. Josephson, *The Politicos* 526 (1938).

Years later a court of appeals observed, "the recurring question which has plagued public regulation of industry [is] whether the regulatory agency is unduly oriented toward the interests of the industry it is designed to regulate, rather than the public interest it is designed to protect." *Moss* v. *CAB*, 139 U. S. App. D. C. 150, 152, 430 F.2d 891, 893. . . .

The Forest Service—one of the federal agencies behind the scheme to despoil Mineral King—has been notorious for its alignment with lumber companies, although its mandate from Congress directs it to consider the various aspects of multiple use in its supervision of the national forests.

The voice of the inanimate object, therefore, should not be stilled. That does not mean that the judiciary takes over the managerial functions from the federal agency. It merely means that before these priceless bits of Americana (such as a valley, an alpine meadow, a river, or a lake) are forever lost or are so transformed as to be reduced to the eventual rubble of our urban environment, the voice of the existing beneficiaries of these environmental wonders should be heard.

Perhaps they will not win. Perhaps the bulldozers of "progress" will plow under all the aesthetic wonders of this beautiful land. That is not the present question. The sole question is, who has standing to be heard?

Those who hike the Appalachian Trail into Sunfish Pond, New Jersey, and camp or sleep there, or run the Allagash in Maine, or climb the Guadalupes in West Texas, or who canoe and portage the Quetico Superior in Minnesota, certainly should have standing to defend those natural wonders before courts or agencies, though they live 3,000 miles away. Those who merely are caught up in environmental news or propaganda and flock to defend these waters or areas may be treated differently. That is why these environmental issues should be tendered by the inanimate object itself. Then there will be assurances that all of the forms of life which it represents will stand before the court—the pileated woodpecker as well as the coyote and bear, the lemmings as well as the trout in the streams. Those inarticulate members of the ecological group cannot speak. But those people who have so frequented the place as to know its values and wonders will be able to speak for the entire ecological community.

Ecology reflects the land ethic; and Aldo Leopold wrote in *A Sand County Almanac* 204 (1949), "The land ethic simply enlarges the boundaries of the community to include soils, waters, plants, and animals, or collectively: the land."

That, as I see it, is the issue of "standing" in the present case and controversy.

JOHN MADDOX

IS CATASTROPHE COMING?

THE DOOMSDAY SYNDROME (1972)

Prophets of doom have multiplied remarkably in the past few years. It used to be commonplace for men to parade city streets with sandwich boards proclaiming "The End of the World Is at Hand!" They have been replaced by a throng of sober people, scientists, philosophers and politicians, proclaiming that there are more subtle calamities just around the corner. The human race, they say, is in danger of strangling itself with pollution, of undermining its essential human character by tampering with heredity and of perverting the whole basis of society with too much prosperity.

The questions which these latter-day doomsday men have raised are subtle and interesting; the spirit in which they are asked is usually too jaundiced for intellectual comfort. Too often, reality is oversimplified or even ignored, so that there is a danger that much of this gloomy foreboding about the immediate future will accomplish the opposite of what its authors intend. Instead of alerting people to important problems, it may seriously undermine the capacity of the human race to look out for its survival. The doomsday syndrome may in itself be as much a hazard as any of the conundrums which society has created for itself.

Nobody should doubt the wish of the contemporary prophets of calamity to find ways out of trouble, and nobody will dispute that modern society is confronted with important tasks which must be urgently tackled. In advanced societies, machinery must be devised for the more equitable treatment of the poor and the disadvantaged. Urban life, although better than it used to be, can be improved. Even where medical care is excellent, ways need still to be discovered of preventing untimely death and unnecessary disease. And in less-developed societies, there are more primitive tasks to be undertaken.

New York: McGraw-Hill, 1972. Excerpts from pages 3–7. Used with permission.

Nourishment is still too fitful. Schooling remains for many a luxury. Housing is shelter for some and a source of envy for others. The whole world knows, after what happened in Bengal in 1971, that Calcutta, for example, presents problems of race, poverty, disease, and freedom that affront and sometimes paralyze the conscience of industrialized societies. The question which the doomsday prophets pose for those who share their compassion for society is whether the energies of the human race should be spent on problems like these which, however difficult, can be solved or whether they should be spent on the avoidance of more distant trouble.

The doomsday cause would be more telling if it were more securely grounded in facts, better informed by a sense of history and an awareness of economics and less cataclysmic in temper. In an age of empiricism, of course, it is always tempting, especially for busy people anxious to get things done, to concentrate on tangible issues and to postpone a consideration of less immediate questions. This, no doubt, is why the doomsday movement has so wonderfully captured the allegiance of the idealistic youth. But is there not an opposing danger that too much preoccupation with the threat of distant calamity may divert attention from good works that might be done? On what horizon should well-intentioned people fix their gaze?

The defect of the case for thinking that calamity is the more important menace is its imprecision. The are some who fear that the burning of fuel on the scale to which modern industry is accustomed will wreck the climate on the surface of the earth, but few meteorologists are able unambiguously to endorse such prophecies. Some fear that the use of pesticides will irrevocably damage the human race, but that is an overdramatic statement of the need carefully to regulate the way in which such chemicals are sprayed on crops. Some fear that modern biology, with genetic processes, will degrade humanity; by doing so, they fly in the face of the past five centuries of the history of medicine, a consistent record of human endeavor. In short, the weakness of the doomsday prophecies is that they are exaggerations. Many of them are irresponsible.

The flavor of the prophecies of disaster is well illustrated by the work of Dr. Paul R. Ehrlich, who startled a good many people with the publication in 1968 of his book *The Population Bomb*. "The battle to feed all of humanity is over," he says. "In the 1970s the world will undergo famines, hundreds of millions of people are going to starve to death in spite of any crash program embarked on now." Dr. Ehrlich goes on to describe in a somber way the rate at which the population of the world is increasing, the inconveniences which are likely to result therefrom and to describe some ways of striking a better balance between population growth and available resources, especially in developing parts of the world. He is against sterilants in drinking water but for compulsory sterilization by other means.

Nobody will deny that it is important to control, if not the size of a population, then its rate of growth. In advanced societies, population control is increasingly a part of good government. In developing countries, it is a prerequisite of economic progress. But famine is now an unreal scarecrow. There is a good chance that the problems of the 1970s and 1980s will not be famine and starvation but, ironically, the problems of how best to dispose of food surpluses in countries where famine has until recently been endemic. Population control is therefore desirable not as a means of avoiding calamity but because like other social benefits—health care, for example—it can accelerate the improvement of the human condition.

Implicit in the fears of the consequences of population growth which are now rife is an oversimple method of prediction. If the population of the world is at present doubling every thirty-five years, does it follow that the population will multiply by four in the next seventy years, so as to reach 14,000 million by the year 2040? In his book *The Population Bomb,* Dr. Ehrlich is scornful about those whom he calls "professional optimists . . . who like to greet every sign of dropping birth rate with wild pronouncements about the end of the population explosion." In his more soberly written *Population, Resources, Environment,* he chooses to base calculations of the future population of the earth on the most pessimistic calculation of the United Nations, which assumes that there will be no change in the fertility of women of childbearing age between now and the end of the century. But in reality, there are already signs that fertility is declining in developing communities in exactly the same way, but possibly more rapidly than fertility declined in Western Europe between fifty and a hundred years ago. One of the strangest features of the Ehrlich description of the population explosion is the bland assumption that developing countries may somehow be unable to follow the path of development along which advanced societies have traveled. Is it any wonder that the predominantly Western preoccupation with the population explosion seems like patronizing neocolonialism to people elsewhere?

In much the spirit in which environmentalists worry about food, unnecessarily as events have shown, so too they worry about natural resources. In the United States, at least, this is an honorable tradition going back to the end of the nineteenth century. Then, Governor Gifford Pinchot was wringing his hands over the prospect that timber in the United States would be used up in roughly thirty years; and that other raw materials such as iron ore and natural gas were rapidly being consumed. Sixty years later, the same complaints are to be heard in the United States. The environmentalists have coined the phrase "our plundered planet" to express the anxiety they feel about the probability that petroleum will be much less plentiful a century from now than it is at present and that the time will come when high-grade copper ores are worked out.

The truth is, however, that society is by no means uniquely dependent on the raw materials now in common use. If copper becomes scarce or merely expensive, aluminum will have to be used instead. If natural diamonds are expensive, then make them synthetically. In any case, although supplies of raw materials like these are known to be limited, the point at which they seem likely to be exhausted recedes with the passage of time so as always to be just over the horizon. Indeed, in spite of what the environmentalists say, the present time appears to be one at which forecasts of scarcity are less valid than ever. Petroleum may be much harder to obtain a century from now, but the past few years have laid the foundations for winning energy from hydrogen and minerals such as uranium in such large quantities that future decades will be much better off than anybody could have expected in the 1950s. And, strange as it may seem, the real economic cost of extracting metals such as lead and copper from the ground is still decreasing as exploration and techniques of mining and metallurgy become more efficient. In economic terms, the earth's resources seem to be becoming more plentiful. . . .

DECLARATION OF THE UNITED NATIONS CONFERENCE ON THE HUMAN ENVIRONMENT

STOCKHOLM, JUNE 5–16, 1972

The United Nations Conference on the Human Environment,

Having met at Stockholm from 5 to 16 June 1972,

Having considered the need for a common outlook and for common principles to inspire and guide the peoples of the world in the preservation and enhancement of the human environment,

I

Proclaims that:

1. Man is both creature and molder of his environment, which gives him physical sustenance and affords him the opportunity for intellectual, moral, social and spiritual growth. In the long and tortuous evolution of the human race on this planet a stage has been reached when, through the rapid acceleration of science and technology, man has acquired the power to transform his environment in countless ways and on an unprecedented scale. Both aspects of man's environment, the natural and the man-made, are essential to his well-being and to the enjoyment of basic human rights—even the right to life itself.

2. The protection and improvement of the human environment is a major issue which affects the well-being of peoples and economic development throughout the world; it is the urgent desire of the peoples of the whole world and the duty of all Governments.

Report of the United Nations Conference on the Human Environment, excerpts from pages 3–4.

3. Man has constantly to sum up experience and go on discovering, inventing, creating and advancing. In our time, man's capability to transform his surroundings, if used wisely, can bring to all peoples the benefits of development and the opportunity to enhance the quality of life. Wrongly or heedlessly applied, the same power can do incalculable harm to human beings and the human environment. We see around us growing evidence of man-made harm in many regions of the earth: dangerous levels of pollution in water, air, earth and living beings; major and undesirable disturbances to the ecological balance of the biosphere; destruction and depletion of irreplaceable resources; and gross deficiencies, harmful to the physical, mental and social health of man, in the man-made environment, particularly in the living and working environment.

4. In the developing countries most of the environmental problems are caused by under-development. Millions continue to live far below the minimum levels required for a decent human existence, deprived of adequate food and clothing, shelter and education, health and sanitation. Therefore, the developing countries must direct their efforts to development, bearing in mind their priorities and the need to safeguard and improve the environment. For the same purpose, the industrialized countries should make efforts to reduce the gap between themselves and the developing countries. In the industrialized countries, environmental problems are generally related to industrialization and technological development.

5. The natural growth of population continuously presents problems for the preservation of the environment, and adequate policies and measures should be adopted, as appropriate, to face these problems. Of all things in the world, people are the most precious. It is the people that propel social progress, create social wealth, develop science and technology and, through their hard work, continuously transform the human environment. Along with social progress and the advance of production, science and technology, the capability of man to improve the environment increases with each passing day.

6. A point has been reached in history when we must shape our actions throughout the world with a more prudent care for their environmental consequences. Through ignorance or indifference we can do massive and irreversible harm to the earthly environment on which our life and well-being depend. Conversely, through fuller knowledge and wiser action, we can achieve for ourselves and our posterity a better life in an environment more in keeping with human needs and hopes. There are broad vistas for the

enhancement of environmental quality and the creation of a good life. What is needed is an enthusiastic but calm state of mind and intense but orderly work. For the purpose of attaining freedom in the world of nature, man must use knowledge to build, in collaboration with nature, a better environment. To defend and improve the human environment for present and future generations has become an imperative goal for mankind—a goal to be pursued together with, and in harmony with, the established and fundamental goals of peace and of worldwide economic and social development.

7. To achieve this environmental goal will demand the acceptance of responsibility by citizens and communities and by enterprises and institutions at every level, all sharing equitably in common efforts. Individuals in all walks of life as well as organizations in many fields, by their values and the sum of their actions, will shape the world environment of the future. Local and national governments will bear the greatest burden for large-scale environmental policy and action within their jurisdictions. International co-operation is also needed in order to raise resources to support the developing countries in carrying out their responsibilities in this field. A growing class of environmental problems, because they are regional or global in extent or because they affect the common international realm, will require extensive co-operation among nations and action by international organizations in the common interest. The Conference calls upon Governments and peoples to exert common efforts for the preservation and improvement of the human environment, for the benefit of all the people and for their posterity.

II

Principles

States the common conviction that:

Principle 1

Man has the fundamental right to freedom, equality and adequate conditions of life, in an environment of a quality that permits a life of dignity and well-being, and he bears a solemn responsibility to protect and improve the environment for present and future generations. In this respect, policies promoting or perpetuating *apartheid*, racial segregation, discrimination, colonial and other forms of oppression and foreign domination stand condemned and must be eliminated.

Principle 2

The natural resources of the earth, including the air, water, land, flora and fauna and especially representative samples of natural ecosystems, must be safeguarded for the benefit of present and future generations through careful planning or management, as appropriate.

Principle 3

The capacity of the earth to produce vital renewable resources must be maintained and, wherever practicable, restored or improved.

Principle 4

Man has a special responsibility to safeguard and wisely manage the heritage of wildlife and its habitat, which are now gravely imperiled by a combination of adverse factors. Nature conservation, including wildlife, must therefore receive importance in planning for economic development.

Principle 5

The non-renewable resources of the earth must be employed in such a way as to guard against the danger of their future exhaustion and to ensure that benefits from such employment are shared by all mankind.

Principle 6

The discharge of toxic substances or of other substances and the release of heat, in such quantities or concentrations as to exceed the capacity of the environment to render them harmless, must be halted in order to ensure that serious or irreversible damage is not inflicted upon ecosystems. The just struggle of the peoples of all countries against pollution should be supported.

Principle 7

States shall take all possible steps to prevent pollution of the seas by substances that are liable to create hazards to human health, to harm living resources and marine life, to damage amenities or to interfere with other legitimate uses of the sea.

Principle 8

Economic and social development is essential for ensuring a favourable living and working environment for man and for creating conditions on earth that are necessary for the improvement of the quality of life.

Principle 9

Environmental deficiencies generated by the conditions of under-development and natural disasters pose grave problems and can best be remedied by accelerated development through the transfer of substantial quantities of financial and technological assistance as a supplement to the domestic effort of the developing countries and such timely assistance as may be required.

Principle 10

For the developing countries, stability of prices and adequate earnings for primary commodities and raw materials are essential to environmental management since economic factors as well as ecological processes must be taken into account. . . .

PART 5

CONTINUATION

A WAVE OF IMPORTANT LEGISLATION, INCLUDING THE CLEAN AIR and Clean Water Acts, and the creation of a visible and active Environmental Protection Agency helped relieve the sense of crisis after 1972, even among some environmental activists. Despite real gains, however, new environmental problems developed and others persisted. Two different energy crises, in 1973 and 1979, forced Americans to question their reliance on imported oil. The discovery of toxic chemicals in the residential neighborhood around the Love Canal in Niagara Falls sparked a national toxics scare in 1978, followed by growing concern about environmental justice and the unequal toxic burdens faced by Americans of different races and classes.

International environmental issues also garnered considerable attention in the 1970s and 1980s. Studies of forests and lakes in northern Europe and North America connected industrial pollution and acidic precipitation to poor forest health and declining aquatic ecosystems. New regulations of sulfur and nitrogen emissions followed. The international community reacted swiftly in the late 1980s, when the discovery of a hole in the ozone layer led to an international ban of chlorofluorocarbons. The international community has been less successful in dealing with global climate change, despite decades of work by concerned scientists and activists.

The documents in this final part of the book give a sense of the continuing environmental struggle, both in the political arena and in individuals' lives. The first is from President Jimmy Carter, who entered the White House during inauspicious times. The American economy had been faltering and, in the wake of the Watergate scandal, cynicism about the federal government had skyrocketed. Carter's "unpleasant talk" about "challenges" facing the nation certainly did not lift the national mood, but his discussion of the need to create a forward-looking energy policy sounds remarkably prescient today.

The second document comes from the congressional debates over the bill that became Superfund, the federal government's response to the realization that toxic dumps and spills threatened Americans well beyond Niagara Falls. Robert A. Roland of the Chemical Manufacturers Association hoped to limit federal involvement, but more important, he hoped to ensure that the legislation would not unfairly burden the chemical industry. In a speech before the House of Representatives that seemed to paraphrase *Pogo*'s declaration nine years earlier, Roland noted that the "problem is societal in scope" and therefore the solution should not be funded through a special tax on chemical companies.

The third document provides a completely different perspective on the toxics scare. This excerpt of an interview with a woman whose family had been relocated from the Love Canal neighborhood reveals the deep emotional scars left by the episode. Although scientists and historians have debated the magnitude of the health threats faced by families who lived around the toxic dump, there can be no debate about the wounds this woman and her family felt.

The final document included in this collection concerns global climate change. It comes from Dr. James E. Hansen, an atmospheric scientist at NASA who became one of the earliest government officials to discuss publicly the human role in global warming. The power of his oral testimony before the Senate, stripped here of the illustrative slides he showed the congressional panel, remains intact. The lingering—indeed perhaps even deepening—debate about global warming, more than twenty years after Hansen's testimony, suggests how much American culture and politics have changed since the environmental moment receded in the early 1970s.

JIMMY CARTER

THE ENERGY PROBLEM: ADDRESS TO THE NATION

SPEECH DELIVERED ON APRIL 18, 1977

Tonight I want to have an unpleasant talk with you about a problem unprecedented in our history. With the exception of preventing war, this is the greatest challenge our country will face during our lifetimes. The energy crisis has not yet overwhelmed us, but it will if we do not act quickly.

It is a problem we will not solve in the next few years, and it is likely to get progressively worse through the rest of this century.

We must not be selfish or timid if we hope to have a decent world for our children and grandchildren.

We simply must balance our demand for energy with our rapidly shrinking resources. By acting now, we can control our future instead of letting the future control us.

Two days from now, I will present my energy proposals to the Congress. Its members will be my partners and they have already given me a great deal of valuable advice. Many of these proposals will be unpopular. Some will cause you to put up with inconveniences and to make sacrifices.

The most important thing about these proposals is that the alternative may be a national catastrophe. Further delay can affect our strength and our power as a nation.

Our decision about energy will test the character of the American people and the ability of the President and the Congress to govern. This difficult effort will be the "moral equivalent of war"—except that we will be uniting our efforts to build and not destroy.

I know that some of you may doubt that we face real energy shortages. The 1973 gasoline lines are gone, and our homes are warm again. But our energy

Excerpts from John T. Woolley and Gerhard Peters, *The American Presidency Project*, Santa Barbara, Calif., available at http://www.presidency.ucsb.edu/ws/?pid=7369

problem is worse tonight than it was in 1973 or a few weeks ago in the dead of winter. It is worse because more waste has occurred, and more time has passed by without our planning for the future. And it will get worse every day until we act.

The oil and natural gas we rely on for 75 percent of our energy are running out. In spite of increased effort, domestic production has been dropping steadily at about 6 percent a year. Imports have doubled in the last five years. Our nation's independence of economic and political action is becoming increasingly constrained. Unless profound changes are made to lower oil consumption, we now believe that early in the 1980s the world will be demanding more oil that it can produce.

The world now uses about 60 million barrels of oil a day and demand increases each year about 5 percent. This means that just to stay even we need the production of a new Texas every year, an Alaskan North Slope every nine months, or a new Saudi Arabia every three years. Obviously, this cannot continue.

We must look back in history to understand our energy problem. Twice in the last several hundred years there has been a transition in the way people use energy.

The first was about 200 years ago, away from wood—which had provided about 90 percent of all fuel—to coal, which was more efficient. This change became the basis of the Industrial Revolution.

The second change took place in this century, with the growing use of oil and natural gas. They were more convenient and cheaper than coal, and the supply seemed to be almost without limit. They made possible the age of automobile and airplane travel. Nearly everyone who is alive today grew up during this age and we have never known anything different.

Because we are now running out of gas and oil, we must prepare quickly for a third change, to strict conservation and to the use of coal and permanent renewable energy sources, like solar power.

The world has not prepared for the future. During the 1950s, people used twice as much oil as during the 1940s. During the 1960s, we used twice as much as during the 1950s. And in each of those decades, more oil was consumed than in all of mankind's previous history.

World consumption of oil is still going up. If it were possible to keep it rising during the 1970s and 1980s by 5 percent a year as it has in the past, we could use up all the proven reserves of oil in the entire world by the end of the next decade.

I know that many of you have suspected that some supplies of oil and gas are being withheld. You may be right, but suspicions about oil companies cannot change the fact that we are running out of petroleum.

All of us have heard about the large oil fields on Alaska's North Slope. In a few years, when the North Slope is producing fully, its total output will be just about equal to two years' increase in our nation's energy demand.

Each new inventory of world oil reserves has been more disturbing than the last. World oil production can probably keep going up for another six or eight years. But some time in the 1980s it can't go up much more. Demand will overtake production. We have no choice about that.

But we do have a choice about how we will spend the next few years. Each American uses the energy equivalent of 60 barrels of oil per person each year. Ours is the most wasteful nation on earth. We waste more energy than we import. With about the same standard of living, we use twice as much energy per person as do other countries like Germany, Japan and Sweden.

One choice is to continue doing what we have been doing before. We can drift along for a few more years.

Our consumption of oil would keep going up every year. Our cars would continue to be too large and inefficient. Three-quarters of them would continue to carry only one person—the driver—while our public transportation system continues to decline. We can delay insulating our houses, and they will continue to lose about 50 percent of their heat in waste. We can continue using scarce oil and natural gas to generate electricity, and continue wasting two-thirds of their fuel value in the process.

If we do not act, then by 1985 we will be using 33 percent more energy than we do today.

We can't substantially increase our domestic production, so we would need to import twice as much oil as we do now. Supplies will be uncertain. The cost will keep going up. Six years ago, we paid $3.7 billion for imported oil. Last year we spent $37 billion—nearly ten times as much—and this year we may spend over $45 billion.

Unless we act, we will spend more than $550 billion for imported oil by 1985—more than $2,500 a year for every man, woman, and child in America. Along with that money we will continue losing American jobs and becoming increasingly vulnerable to supply interruptions.

Now we have a choice. But if we wait, we will live in fear of embargoes. We could endanger our freedom as a sovereign nation to act in foreign affairs. Within ten years we would not be able to import enough oil—from any country, at any acceptable price.

If we wait, and do not act, then our factories will not be able to keep our people on the job with reduced supplies of fuel. Too few of our utilities will have switched to coal, our most abundant energy source.

We will not be ready to keep our transportation system running with smaller, more efficient cars and a better network of buses, trains and public transportation.

We will feel mounting pressure to plunder the environment. We will have a crash program to build more nuclear plants, strip-mine and burn more coal, and drill more offshore wells than we will need if we begin to conserve now. Inflation will soar, production will go down, people will lose their jobs. Intense competition will build up among nations and among the different regions within our own country.

If we fail to act soon, we will face an economic, social and political crisis that will threaten our free institutions.

But we still have another choice. We can begin to prepare right now. We can decide to act while there is time.

That is the concept of the energy policy we will present on Wednesday. Our national energy plan is based on ten fundamental principles.

The first principle is that we can have an effective and comprehensive energy policy only if the government takes responsibility for it and if the people understand the seriousness of the challenge and are willing to make sacrifices.

The second principle is that healthy economic growth must continue. Only by saving energy can we maintain our standard of living and keep our people at work. An effective conservation program will create hundreds of thousands of new jobs.

The third principle is that we must protect the environment. Our energy problems have the same cause as our environmental problems—wasteful use of resources. Conservation helps us solve both at once.

The fourth principle is that we must reduce our vulnerability to potentially devastating embargoes. We can protect ourselves from uncertain supplies by reducing our demand for oil, making the most of our abundant resources such as coal, and developing a strategic petroleum reserve.

The fifth principle is that we must be fair. Our solutions must ask equal sacrifices from every region, every class of people, every interest group. Industry will have to do its part to conserve, just as the consumers will. The energy producers deserve fair treatment, but we will not let the oil companies profiteer.

The sixth principle, and the cornerstone of our policy, is to reduce the demand through conservation. Our emphasis on conservation is a clear difference between this plan and others which merely encouraged crash production efforts. Conservation is the quickest, cheapest, most practical source of energy. Conservation is the only way we can buy a barrel of oil for a few dollars. It costs about $13 to waste it.

The seventh principle is that prices should generally reflect the true replacement costs of energy. We are only cheating ourselves if we make energy artificially cheap and use more than we can really afford.

The eighth principle is that government policies must be predictable and certain. Both consumers and producers need policies they can count on so they can plan ahead. This is one reason I am working with the Congress to create a new Department of Energy, to replace more than 50 different agencies that now have some control over energy.

The ninth principle is that we must conserve the fuels that are scarcest and make the most of those that are more plentiful. We can't continue to use oil and gas for 75 percent of our consumption when they make up seven percent of our domestic reserves. We need to shift to plentiful coal while taking care to protect the environment, and to apply stricter safety standards to nuclear energy.

The tenth principle is that we must start now to develop the new, unconventional sources of energy we will rely on in the next century. . . .

Other generations of Americans have faced and mastered great challenges. I have faith that meeting this challenge will make our own lives even richer. If you will join me so that we can work together with patriotism and courage, we will again prove that our great Nation can lead the world into an age of peace, independence, and freedom.

Thank you very much, and good night.

ROBERT A. ROLAND

STATEMENT REGARDING SUPERFUND

OCTOBER 10, 1979

My name is Robert A. Roland, President of the Chemical Manufacturers Association (CMA). Today I am speaking on behalf of CMA, a nonprofit trade association having 190 company members that represent more than 90 percent of the production capacity of basic industrial chemicals within this country.

The members of our Association are acutely aware that some substances, when spilled or improperly disposed of, can pose a significant threat to human health and the environment. We are particularly concerned about the problems caused by certain waste disposal sites. Two of the most widely publicized examples of failing sites are New York's Love Canal and Kentucky's Valley of the Drums. We believe it is essential that sites which present an imminent hazard be dealt with in a prompt and effective manner. Steps should be taken to quantify the problems, mitigate the damages and control the hazard. Inaction or needless delay is not acceptable.

Your Subcommittee, as these hearings demonstrate, recognizes the importance of developing mechanisms to deal with such problems. One approach that is receiving active consideration is the creation of a Federal relief fund, often described as a "Superfund" or "ULTRAFUND."

Some have proposed that such a fund cover a broad range of incidents, including oil spills, hazardous substance spills and abandoned and inactive hazardous waste disposal sites. This would be a serious mistake. . . .

Public funding is appropriate for reasons of equity and sound public policy.

House Subcommittee on Transportation and Commerce, *Superfund: Hearings on H.R. 4571, H.R. 4566, and H.R. 5290*, 96th Cong., 1st sess., 1979, excerpts from pages 351–52, 356–59, 363–65, and 376–77.

It is clearly inequitable to place the burden on today's companies, stock-holders, or customers for the practices, failures or shortcomings of yesterday's industrial producers. The generators of today's hazardous wastes, many of whom are handling their waste in acceptable ways, must not bear the cost of past failures.

Furthermore, many of the problems of today were not known and were, in fact, unknowable in terms of the state of knowledge and art at the time of the original disposal. Companies must not be assessed now for disposal practices that were considered acceptable at the time.

A *Washington Post* editorial of June 23, 1979, affirms this important principle. In exploring the question of who should pay the costs of cleanup, the editorial points out that punishing the polluter makes sense where someone has been negligent but is not equitable in other cases. According to the *Post*:

> Punishing the polluter is appropriate where someone has been negligent—for instance, by dumping toxic wastes illegally or by failing to report health threats that later appear. Yet this course is not effective in the worst cases, where the polluter cannot be found or cannot pay the bills. And it is not equitable in those frustrating situations where the waste disposal was done in accord with the laws of the time. In such cases, out of necessity or fairness, the public will have to bear much of the cost.

A similar point has also been made by Congressman Bob Eckhardt (D.-TX, -8), Chairman of the House Commerce Subcommittee on Oversight and Investigation. In response to a question by Bob Abernathy on the *Today Show*, Eckhardt stated that if a company wasn't negligent then the site should be treated like a hurricane and cleaned up at public expense. Following is the dialogue between Mr. Abernathy and Congressman Eckhardt:

> ABERNATHY: But the stuff that's already there—suppose the company wasn't negligent; it just didn't know any better. Or suppose the company has gone out of business—who cleans up then? Who pays?

> ECKHARDT: Some of this was put down way back in World War I, so we've got to work out some way to clean it up at public expense, just as we would take care of any other tragic situation which might occur. Like, for instance, the effects of a hurricane.

Hazardous wastes are not the product of chemical industry activities alone. Rather, they are an integral by-product of our industrial society, and the by-products of the daily life of every citizen. The problems associated with abandoned hazardous waste disposal sites reflect more than 100 years of industrial development in the Nation. Hazardous wastes have been and will continue to be generated by a wide range of industries, business concerns, government agencies and defense installations, municipalities, and scientific facilities. The problem is societal in scope and the mechanisms for coping with it should reflect its societal nature, just as the benefits of resolving the problems will inure to all elements of the country. Only through the use of public monies, both state and Federal, can this general responsibility be fairly discharged.

In addition, the use of Federal and state revenues would be of significant benefit in assuring fiscal and operational accountability and responsibility. This avoids the pitfalls of trust-fund financing, where the fund is self-perpetuating and it becomes impossible to balance the level of funding with the needs of society once the fund has been created. Otherwise there would be a clear tendency to overspend—wasting resources—since someone else's money would be involved. The interest of sound public policy and fiscal control dictate public revenues for this reason alone.

In addition, the Administration bill proposes to fund the program largely through a tax or fee imposed on the chemical industry. Such a chemical industry fee, levied on petrochemical feedstocks and certain basic inorganic chemicals, would be highly inequitable. EPA has itself identified 17 industries whose operations result in generation of hazardous waste. It would bear no relationship to responsibility for past wrongful dumping, and no relationship to the scope of individual containment efforts.

As discussed later, in singling out the chemical industry for punishment in the fee system, the Administration bill is like a bill of attainder forbidden under the Constitution. In imposing a fee without regard to any benefit derived or the discharge of any responsibility, the Administration's approach will encounter additional Constitutional difficulties. . . .

WHAT ELSE SHOULD BE DONE TO ADDRESS DISPOSAL SITE PROBLEMS, PRESENT AND FUTURE?

Besides the creation of a Federal emergency cleanup fund for old dumpsites, CMA believes that a range of additional actions are needed. We recommend the following:

1. Inventory disposal sites—state by state—to identify more accurately the: location; number; and potential dangers from old sites. We believe that the EPA data (the Hart study) are so incomplete, imprecise and poorly structured that there can be little confidence in resulting conclusions about the true scope of the problem, the real degree of hazard, and the actual costs of remedial action needed.

At a minimum, a serious effort must be undertaken to fully understand the dimensions of the problems before moving ahead with a massive ULTRA-FUND. This is not to argue for delay in addressing emergency situations as they arise, but merely an intelligent attempt to quantify and direct the responses toward truly hazardous conditions. States are the most appropriate governmental entities to conduct these studies. They have the best first-hand knowledge of activities that occurred within their own borders. For instance, an interim report has been recently released giving results of a survey of disposal sites within the State of New York:

Hazardous Sites	520
Industry-owned Sites	380
Government-owned Sites	62
Privately-owned Sites	69
No Known Owner	9
Sites Undergoing Current Remedial Action	24
Potentials for Additional "Love Canals"	0

The State Environmental Conservation Commissioner was quoted as saying that the interim report "expands" work begun as a result of the Love Canal incident, but that "nothing that we have seen approaches the magnitude of the Love Canal problem."

The New York survey, if reflective of the national situation as a whole, does not seem to justify the inflammatory "ticking time bomb" rhetoric so often used to characterize disposal site problems by those seeking the ULTRAFUND panacea. . . .

SUMMARY

In summary, we contend that the case for an ULTRAFUND lacks a sound data base, probably overstates both health and environmental dangers, and probably overestimates the degree of remedial action needed and the costs of such action. We believe the ULTRAFUND approach is unfair, administratively unworkable, poorly justified, economically unsound, and very likely unconstitutional.

We urge your Subcommittee to focus on the problem of "orphan" dumpsites—the area which clearly requires prompt legislative action. The chemical industry will continue to offer constructive cooperation in the effort to develop appropriate solutions.

A MOTHER'S REFLECTIONS
ON THE LOVE CANAL DISASTER

OCTOBER 1982

This couple lived in Ring III for about twelve years before they relocated. Both are in their middle years; they have two small children. The husband is a college graduate, employed in the public sector. His wife works out of their home. They believe contamination is widespread in the area.

My husband remembers when that canal was an open canal. He used to swim in it. He said his mother used to tell them, when they came home, they smelled like the sewer. . . . But I wasn't aware, I didn't know what the Love Canal was. I lived there for 12 years and I'd never heard of it. That spring, I saw some government cars and people there on Frontier Avenue. We used to ride bikes through the area and you could smell it. And I used to wonder what it was you could smell. It was a very distinct chemical smell but, heck, that was three or four blocks away. It didn't bother me. I didn't smell it at my house so I never got particularly concerned. I was just curious as to why there were U.S. Government cars over there and some people with those survey machines, equipment, and so on. . . .

It wasn't until the summer of '78 that I was really aware there was a problem. When that canal thing hit, August first, I went over to my neighbor and said, "Hey, I don't know what this whole canal thing is about, but you can bet it's going to affect our property value. There's a meeting tonight, why don't you go?" He said, "We lived here for 40 years, don't get worried. There's no prob-

---

As found in Martha R. Fowlkes and Patricia Y. Miller, *Love Canal: The Social Construction of Disaster,* a Report for the Federal Emergency Management Agency, excerpts from pages 90–95.

lem." I said, "I don't know if you folks are aware, but we're expecting another baby and our older one has a lot of birth defects. And I don't like the sound of them moving pregnant mothers out four blocks away. We're going to get involved. . . . "

As soon as we heard that they wanted expectant mothers and children under two removed in Rings I and II—because of the high birth defect rate, because of the risk to the fetus, because of the excessive amount of kidney disorders, heart disorders, digestive disorders—we thought, "Oh God, that's exactly what's wrong with our son, heart, kidneys, pancreas and bladder." All of a sudden these things are hitting you off the front of the newspaper and they're talking about your neighborhood. I said, "I don't believe this. It makes sense, that's what's wrong with him."

He had a urinary tract obstruction that's been repaired. He had another kidney surgery recently. Originally, it was a birth defect, and the second time they repaired this, it was because there was an obstruction in the urinary tract, scar tissue. And they had to repair it again for a third time. He has a heart murmur. When he was a pre-schooler, finally when he could talk, he would tell you, "Mommy, my tummy hurts." He'd come in from playing, he'd be white as a ghost, doubling over with pain, and his stomach would bloat up like a balloon. The doctor said it didn't have anything to do with his kidneys. He sent us to the gastroenterology clinic. They did so many tests on him. Finally they said, "He's doing this to get attention." I just hit the roof. I said, "A kid cannot turn pale to get attention. He does not bloat up to get attention. He's an only child. I do not work, he is not lacking for attention. There's something wrong with this boy." They put him back in. He was in every two or three weeks. He was doubling over, he'd wake up crying that his stomach hurt. Finally, they determined that his pancreas did not produce enough enzymes. He didn't produce enough to digest the lactose and sucrose in the diet, and the undigested sugars would turn to gas, which would cause him to bloat, which caused the distress. His waist would be two inches smaller in the morning than at night. . . .

It was a nerve disorder of the pancreas. It's something he has outgrown to a degree. He could not tolerate any milk or sugar in his diet and was on soybean milk until about a year ago. . . . When he was a baby, he would throw up all the time. I really think that part of his fussiness as an infant was because of his intolerance to milk. It was not recognized as such and, I think, today that's why he gets allergies to a lot of things. He can't breathe worth a darn. . . . He has allergies to pollution, air pollution, house dust.

He seemed to be the sickest kid around. We were always broke with doctors' bills. We thought all kids were sick. But we found out, after we got out of there, how much our doctor bills decreased. Our kid was sick so you didn't think anything of it, except that when I would talk about spending a hundred dollars a month on office calls and prescriptions, my girlfriends would say they didn't spend a hundred dollars a year. I couldn't believe it. I believe it now. . . .

We had no idea when we bought the house. When they tested our house for chemical readings in the basement, the first time they tested it, they sent back these little computer forms giving us names and numbers. It didn't mean anything to me. I couldn't pronounce the names, and I didn't have anything to compare the numbers to. Were they on a scale of one to ten? One to a million? It didn't mean a thing. So I called a person who had a doctorate in chemistry and was in charge of an industrial chemistry lab. I called him up and said, "If I give you the names of some chemicals and some numbers, can you tell me what they mean?" He said, "Well, give me a try." I spelled off these words to him, chloroform, trichloroethylene. I can't remember all the words, there were about five or six of them. He said, "Why?" I said, "Well, you've probably heard of the Love Canal. These were the readings they came up with in our basement." He said, "Can you smell anything down there?" I said, "No." He said, "All right, I'll call you back." So he called me back a couple of days later and he said, "My God, your chloroform reading is so high they wouldn't allow it under OSHA standards in the chem lab. That chloroform would be allowed only in a pass-thru area." I said, "That's in my basement. I'm three months pregnant and I've got a kid with multiple birth defects and we're fixing up the basement for a playroom." He said, "My God, how can the state tell you to stay there and you're pregnant. Get the hell out of your house. I wouldn't spend any time in that basement. And the only thing between that basement and your kitchen floor and your living room floor is about two inches of wood. If I were you, I'd fight to get out of there. . . . "

We never smelled a thing, never. That's the thing that's so frightening, that you never did smell anything. But when you think about it, chloroform or natural gas is not detectable either, so it did make sense in those terms. Then another thing, when they came back—we made a stink after we found out what those readings meant, us and some other families—they retested and all our readings dropped to zero. I say, "Isn't that interesting? Now who do I believe? Do I believe a friend who is telling me, who has nothing to lose one way or the other, or do I believe the state that all of a sudden my chemical readings dropped to zero?" I said, "I want out. I'm not staying here to find out. . . . "

There was no doubt in my mind that I wanted out of there as soon as I heard that somebody somewhere along the line had determined there was a high risk of birth defects and danger to the fetus, and I was pregnant. I knew what my child had. I wanted out. I liked my house, but I didn't like it to the degree that I was going to stay there and fight for it. I wanted to get out of there, because, to me, you couldn't put a price on your child's health. I had seen what happened to my son and I couldn't prove one way or the other that it was a direct result of the Love Canal, but I was convinced enough that it was not worth staying there to find out. Whatever the price was, we had to get out of there. . . .

We were very active. We lost friendships with some people who didn't feel there was any problem there and that you were just doing it to sell your house, or to get out of there. I personally don't care, because I feel that I was right. I will never change my mind. I feel that it is a health hazard. I would not wish anybody to go through what we went through. You couldn't make plans, you couldn't do anything. You were waiting to find out if somebody was going to buy your house. . . . And I know that Lois Gibbs was criticized by many people but I give her a lot of credit. She really personally got everybody out. . . .

I really resent the EPA not coming out and making some kind of government statement. It's disgraceful the way they've handled that. A lot of people, I think, wanted to stay there. Fine. They're adults, they're intelligent people. They have two years to think about it, read about it, see what's going on around them. But I don't think that the EPA has helped at all by not having a reasonably prompt report. They really seem to be backing off, waiting and waiting to soften the blow, or downplay it, or what. You always had the feeling that they were spoon-feeding you information. They'd come out with a little information and everybody goes all crazy. Then everything kind of dies down. And they come out with another little bit of spoon-fed information—"no significant level," this has happened, such and such has been found, chromosome damage in people and so on, but there's "no significant level." They always quote this "no significant level" until you'd like to throw it back at them, you're so sick of hearing that. And always on Friday afternoons, so there's no government agencies open after five o'clock. . . .

It's been very disappointing the way the information has been disseminated, how it varied from agency to agency, tests. We are very skeptical about any of their test results now because I don't think they've been completely honest. I think that a lot of times they give you half truths. Like having them read the house and saying it was a low level and then finding out from another source it was not a low level. After that, I don't care what they tell me. I don't care what they tell me about the health effects of Love Canal. I'm convinced that it did affect my son and will affect him and will affect me for the rest of my life. Physically, emotionally, mentally, financially. The total impact of the thing is something you will never recover from.

DR. JAMES E. HANSEN

TESTIMONY REGARDING THE GREENHOUSE EFFECT AND GLOBAL CLIMATE CHANGE

Senator Wirth and Senator Murkowski, thank you for the opportunity for me to testify. Before I begin, I would like to state that although I direct the NASA/ Goddard Institute for Space Studies, I am appearing here as a private citizen on the basis of my scientific credentials.

The views that I present are not meant to represent in any way agency or administration policy. My scientific credentials include more than 10 years' experience in terrestrial climate studies and more than 10 years' experience in the exploration and study of other planetary atmospheres.

I will summarize the result of numerical situations of the greenhouse effect, carried out with colleagues at the Goddard Institute. Previous climate modeling studies at other laboratories and at our own examined the case of doubled carbon dioxide, which is relevant to perhaps the middle of the next century based on expected use of fossil fuels.

The unique aspect of our current studies is that we let CO_2 and other trace gases increase year by year as they have been observed in the past 30 years, and as projected in the next 30 years. This allows us to predict how climate will change in the near term, and to examine the question of when the greenhouse effect will be apparent to the man in the street.

We began our climate simulation in 1958, when CO_2 began to be measured accurately, as shown on the first viewgraph. Measurements of other trace gases such as methane, chlorofluorocarbons and nitrous oxide began more recently, but their trends can be estimated with reasonable accuracy back to 1958.

For the future, it is difficult to predict reliably how trace gases will con-

Senate Committee on Energy and Natural Resources, *Greenhouse Effect and Global Climate Change: Hearings, 100th Cong., 1st sess., 1987*, excerpts from pages 51–54.

tinue to change. In fact, it would be useful to know the climatic consequences of alternative scenarios. So we have considered three scenarios for future trace gas growth, shown on the next viewgraph.

Scenario A assumes the CO_2 emissions will grow 1.5 percent per year and that CFC emissions will grow 3 percent per year. Scenario B assumes constant future emissions. If populations increase, Scenario B requires emissions per capita to decrease.

Scenario C has drastic cuts in emissions by the year 2000, with CFC emissions eliminated entirely and other trace gas emissions reduced to a level where they just balance their sinks.

These scenarios are designed specifically to cover a very broad range of cases. If I were forced to choose one of these as most plausible, I would say Scenario B. My guess is that the world is now probably following a course that will take it somewhere between A and B.

We have used these three scenarios in our global climate model, which simulates the global distribution of temperatures, winds, and other climate parameters. Running our model from 1958 to the year 2030, the results for the global mean temperature are as shown in the next viewgraph. The model yields warming by a few tenths of a degree between 1958 and today. In fact, the real world shown by the black curve, has warmed by something of that order.

This warming is not large enough relative to the natural variability of climate for us to claim that it represents confirmation of the model. But we may not have long to wait [for] warming of 0.04 of a degree centigrade, which is three times the standard deviation of the natural variability of the global temperature; if that is maintained for several years, that will represent strong evidence that the greenhouse effect is on this track.

If the world follows trace gas Scenario A or B or something in between, the model says that within 20 years global mean temperature will rise above the levels of the last two interglacial periods and the earth will be warmer than it has been in the past few hundred thousand years.

The man in the street is not too concerned about global mean, annual mean temperature, so let us look at maps of the predicted temperature change for a particular month.

The next viewgraph shows the computer temperature anomalies for July for the intermediate Scenario B. This shows July 1986 in the upper left, going to July of 1987, 1990, and then on the right, 2000, 2015, and 2029. The yellows and reds are the areas that are significantly warmer than the 1950's climatology. Blues are areas that are colder than normal.

The map, for any given month, represents natural fluctuations or noise of the climate systems as well as a long-term trend due to the greenhouse effect.

The natural fluctuations are an unpredictable sloshing around of a nonlinear fluid dynamical system. So these maps should not be taken as predictions of the precise patterns for a particular year.

One conclusion that I want to draw from these maps is that at the present time in the 1980's in a given month, there are almost as many areas colder than normal as areas warmer than normal. This is because the greenhouse warming is smaller than the natural fluctuation of regional climate.

You can see that by 13 years from now, the year 2000, the probability of being warmer than normal is much greater than being cooler than normal. In a few decades from now, it is warm almost everywhere.

So how important are temperature anomalies of this magnitude? One indication is provided by recent experience in the real world. The next viewgraph shows observational data for July 1986 on the top and July 1987 on the bottom.

You probably remember that in July of 1986 there was a heat wave in the Southeast United States, and in July of 1987 it was warm on the East Coast. The same color scale is used here as for the model results, so you can use the last two Julys in the real world to gage the yellow and red colors in the preceding maps.

This makes it obvious that the model predictions for the future shown on the earlier graph represent a major increase in the frequency and severity of July heat waves.

In the letter requesting my testimony, you asked me specifically to address the question of how the greenhouse effect may modify the temperature in the Nation's city. The next viewgraph shows estimates of the number of days per year in which the temperature exceeds a given threshold.

For example, for Washington, D.C., the number of days in which the temperature exceeds 100 degrees Fahrenheit has been one day per year on the average in the period 1950 to 1983. In the doubled CO_2 climate, which will be relevant to the middle of the next century if the world follows trace gas Scenario A, there are about 12 days per year above 100 degrees Fahrenheit.

The number of days per year with temperature exceeding 90 degrees Fahrenheit increases from about 35 to 85, and the number of nights in which the minimum does not drop below 80 degrees Fahrenheit increases from less than one per year to about 20 per year in our climate model.

Obviously, if the greenhouse effect develops to this extent, it will have major impacts on people. The doubled CO_2 level of climate change is not expected until, perhaps, the middle of the next century. It is difficult to predict when, because it depends upon which emission scenario the world follows.

Predictions for the more immediate future are shown on the next view-graph. This shows the number of days with temperatures exceeding 90 degrees Fahrenheit decade by decade. Climatology is the 1950's through the 1970's. The results for Scenario A, which we described as business as usual are on the left.

Scenario C on the right has drastic and probably implausible emission cuts. The conclusion that I draw from this graph is these climate impacts depend greatly on the emission scenario which the world follows.

The climate impacts in Scenario A become dramatic by the 2030s, but in Scenario C the main effect remains smaller than the year to year natural variability.

Finally, I would like to comment on an obvious question: How good are these climate predictions? The climate models we employ and our understanding of the greenhouse effect have been extensively tested by simulations of a range of climates which existed at past times on the earth and on other planets.

So we know the capabilities and limitations of the global models reasonably well. There is, in fact, a substantial range of uncertainty in the predicted temperature change. For example, we can only say that the global climate sensitivity, the doubled CO_2, is somewhere in the range from 2 degrees centigrade to 5 degrees centigrade.

The model used in our studies has a sensitivity of 4 degrees centigrade, which is in the middle of the range obtained from other global climate models.

The geographical patterns of greenhouse climate effects are uncertain, especially changes in precipitation, as Dr. Manube will discuss. However, the uncertainties in the nature and patterns of climate effects cannot be used as a basis for claiming that there may not be large climate changes.

The scientific evidence for the greenhouse effect is overwhelming. The greenhouse effect is real, it is coming soon, and it will have major effects on all peoples. As greenhouse effects become apparent, people are going to ask practical questions and want quantitative answers. Before we can provide climate projections with the specificity and the precision that everyone would like, we first must have major improvements in our observations and understanding of the climate system.

In my submitted testimony, I have listed observations which I believe are most crucial. I believe it is very important that observational systems be in place by the 1990's as greenhouse effects become significant. That is necessary if we are able to provide to decision-makers improved information as the greenhouse effect grows, and as its importance to society [increases].

Thank you for this opportunity to express my opinion.

BIBLIOGRAPHICAL ESSAY

PRIMARY READINGS

A raft of environmental literature appeared in the years from 1968 to 1972, and these primary sources provide wonderful insight into the issues that drove environmental activism. Just as important, they reveal the urgent tenor of the crisis. I encourage interested students to seek out these works, because most of them are easily found and very readable.

Some of the works published during the environmental moment became classics, including Paul R. Ehrlich's *The Population Bomb* (New York: Ballantine Books, 1968) and Barry Commoner's *The Closing Circle: Nature, Man, and Technology* (New York: Knopf, 1971). Other central texts in the environmental movement appeared during these years, such as the Club of Rome's assessment of the environmental crisis, published as *The Limits to Growth* (New York: Universe Books, 1972). Charles A. Reich's early assessment of the movement, *The Greening of America* (New York: Random House, 1970), is very instructive. *Ecotactics: The Sierra Club Handbook for Environmental Activists*, first published in 1970, also provides a valuable window into activists' thoughts and actions.

Several other works from this period have had lasting influence, including E. F. Schumacher's *Small Is Beautiful: Economics As If People Mattered* (New York: Harper & Row, 1973); Paul Swatek's *The User's Guide to the Protection of the Environment: The Indispensable Guide to Making Every Purchase Count* (New York: Friends of the Earth/Ballantine, 1970); and Barbara Ward and René Dubos, *Only One Earth: The Care and Maintenance of a Small Planet* (New York: Norton, 1972). For an introduction to the first Earth Day, an excellent place to start is *Earth Day, the Beginning: A Guide to Survival Compiled and Edited by the National Staff of Environmental Action* (New York: Bantam Books, 1970).

Many lesser-known works fill out the long list of environmental publications during the environmental moment, including the work of dozens of periodicals that put together environmentally themed special issues. Among the most important was the *Daedalus* special issue, published in the fall of 1967, called "America's Changing Environment," which was republished in 1970 with the same title by Beacon Press, with a new introduction by Roger Revelle and Hans H. Landsberg. Two years later, Houghton Mifflin republished the special issue of *The Ecologist* (January 1972), authored by Edward Goldsmith, Robert Allen, and others, under the title *Blueprint for Survival*. Even *Fortune* published its environmental reportage in paperback under the title *The Environment: A National Mission for the Seventies* (New York: Harper & Row, 1970). Interested students might also locate the individual special issues of popular magazines, including *National Geographic*'s "Our Ecological Crisis" issue from December of 1970, or perhaps the more radical *Ramparts* ecology special (May 1970).

In preparation for the Stockholm Conference, Massachusetts Institute of Technology faculty, with the aid of many other scientists, initiated a Study of Critical Environmental Problems in 1969. The involved scientists met in the summer of 1970, and the results of their work appeared in print soon thereafter under the title *Man's Impact on the Global Environment: Assessment and Recommendations for Action* (Cambridge, Mass.: MIT Press, 1970). The United Nations itself published documents related to the Stockholm Conference, including *Man's Home* (1972), a nicely packaged collection of pamphlets on environmental topics, including urbanization, pollution, resource depletion, and development.

SECONDARY READINGS

Scholars have produced a great deal of work on environmental thought, policy, and activism, but until recently surprisingly little of it focused primarily on the environmental moment. I cannot pretend that this essay is comprehensive, but it does describe the literature that has been most helpful to me as I shaped this collection and my own understanding of environmentalism in the late 1960s and early 1970s.

Samuel P. Hays wrote what has long been the standard text on environmentalism, with a traditional approach and a heavy emphasis on politics, but his *Beauty, Health, and Permanence: Environmental Politics in the United States, 1955-1985* (New York: Cambridge University Press, 1987) is not an easy read. A better place to start might be Hal K. Rothman's *The Greening of a Nation?: Environmentalism in the United States since 1945* (New York: Harcourt Brace

College Publishers, 1998), which offers a good overview of the long postwar era, with a special emphasis on the West, wilderness, and national parks. I also like Philip Shabecoff's *A Fierce Green Fire: The American Environmental Movement* (Washington, D.C.: Island Press, 2003), which is very readable and full of useful information. Finally, wildlife biologist Victor B. Scheffer wrote a valuable survey of environmental issues in the postwar era in *The Shaping of Environmentalism in America* (Seattle: University of Washington Press, 1991).

Of the books with a broad approach to the movement, Robert Gottlieb's *Forcing the Spring: The Transformation of the American Environmental Movement* (Washington, D.C.: Island Press, 1993) is perhaps the most influential. Gottlieb convinced many scholars to look beyond the mainstream environmental organizations to explain and describe environmentalism, and since then historians have paid much more attention to grassroots activism, especially that led by women and minorities.

More recently, Adam Rome has made significant contributions to our understanding of the environmental movement, most importantly with *The Bulldozer in the Countryside: Suburban Sprawl and the Rise of American Environmentalism* (New York: Cambridge University Press, 2001). This scholarly yet accessible book links suburban environmental problems with growing environmental activism. Rome also published an important article that placed the environmental moment in the context of 1960s political culture: "'Give Earth a Chance': The Environmental Movement and the Sixties," *Journal of American History* (September 2003): 525-54.

Many works provide a window on the environmental moment through the study of a particular place or issue. Among the most important of these is Andrew Hurley's *Environmental Inequalities: Class, Race, and Industrial Pollution in Gary, Indiana, 1945-1980* (Chapel Hill: University of North Carolina Press, 1995), which describes how environmental regulation actually increased poor and Black communities' exposure to certain pollutants, even as conditions for wealthier residents of northern Indiana improved. See also Chad Montrie, *To Save the Land and People: A History of Opposition to Surface Coal Mining in Appalachia* (Chapel Hill: University of North Carolina Press, 2003), which provides an excellent history of Appalachia's previously understudied struggle to regain control of its own land and stop the damage of strip mining in the 1960s and 1970s. For an excellent case study of how environmentalism influenced an urban community, see Jeffrey Sanders, *Seattle and the Roots of Urban Sustainability: Inventing Ecotopia* (Pittsburgh: University of Pittsburgh Press, 2010). James Longhurst's *Citizen Environmentalists* (Medford, Mass.:

Tufts University Press, 2010) makes a broader point about the role of citizen activism in shaping the environmental movement while providing a useful case study of Pittsburgh's struggle with air pollution during the environmental moment.

Several biographies of key players in the environmental moment also provide insight into the larger movement. See Michael Egan's fine biography *Barry Commoner and the Science of Survival: The Remaking of American Environmentalism* (Cambridge: MIT Press, 2007) and Andrew G. Kirk's *Counterculture Green: The Whole Earth Catalog and American Environmentalism* (Lawrence: University Press of Kansas, 2007), which offers an excellent discussion of Stewart Brand and the appropriate technology movement's contribution to the environmental moment. Interested readers may also find political biographies of use, including J. Brooks Flippen, *Nixon and the Environment* (Albuquerque: University of New Mexico Press, 2000); and Adam M. Sowards, *The Environmental Justice: William O. Douglas and American Conservation* (Corvallis: Oregon State University Press, 2009), which offers a brief biography of one of the most important legal minds of the movement.

I have found several other books useful in thinking about the development of the environmental crisis. On the relationship between film and nature appreciation, see Gregg Mitman, *Reel Nature: America's Romance with Wildlife on Film* (1999; rpt., Seattle: University of Washington Press, 2009). Like Mitman, Mark V. Barrow Jr. studies a much longer period than just the environmental moment, but his *Nature's Ghosts: Confronting Extinction from the Age of Jefferson to the Age of Ecology* (Chicago: University of Chicago Press, 2009) offers a thorough telling of the effort to protect individual species. This bibliographical essay cannot contain all the valuable articles written about the environmental moment, but because I found it so helpful I will mention Finis Dunaway's "Gas Masks, Pogo, and the Ecological Indian: Earth Day and the Visual Politics of American Environmentalism," *American Quarterly* (March 2008): 67-99.

Anyone interested in understanding the development of American environmentalism must study the role of Rachel Carson. Start with *Silent Spring*, published first in 1962 and many times since, and then read Linda Lear's comprehensive biography, *Rachel Carson: Witness for Nature* (New York: Henry Holt and Company, 1997). For a broader context regarding pesticides, see Thomas Dunlap's *DDT, Silent Spring, and the Rise of Environmentalism* (Seattle: University of Washington Press, 2008).

The scholarly examination of the environmental moment is still very much in its infancy, and important books appear every year. Much more work needs to be done on the international aspect of the environmental moment, but interested students can find valuable works on environmentalism outside the United States. See, for example, Jeffrey Broadbent, *Environmental Politics in Japan: Networks of Power and Protest* (New York: Cambridge University Press, 1998); and Russell J. Dalton, *The Green Rainbow: Environmental Groups in Western Europe* (New Haven, Conn.: Yale University Press, 1994).

INDEX

National Environmental Policy Act
(1969), 59, 64–67, 126
National Geographic, 6
National Review, 59–60, 68–69
Native Americans, 56, 60, 77–79
Nature, 108
Nelson, Gaylord, 61; Earth Day
Speech, 85–86
Newfield, Jack: "Lead Poisoning:
Silent Epidemic in the Slums," 36,
43–48
New York City, 43–48
Nile River, 40
Nixon, Richard, 4, 43, 60, 69; "Special
Message to the Congress on Envi-
ronmental Quality," 71–76
noise, 21, 39, 50, 58, 88
nuclear: energy, 132, 144–45; testing,
13, 114; war, 31–32; weapons, 6, 59.
See also strontium 90

oil. *See* petroleum
Ohio, 94. *See also* Cleveland
Oregon, 62–63, 97–99
Osrin, Ray, 62: "Someday Son, All This
Will Be Yours," 96
overcrowding, 57, 61, 81, 85, 88, 91. *See
also* population

passenger pigeon, 39
penicillin, 18
petroleum, 21, 70, 115–16, 132, 142–43;
spills, 36, 54–55, 116, 146
Phinney, Eleanor, 62; "Letter to the
Oregon Environmental Council,"
97–99
Pinchot, Gifford, 131
Pittsburgh, 8, 13, 49–51, 63, 100–101
planning, 62, 66–67, 97–99, 136, 142
Pogo, 61, 84, 140

pollution, 3, 20–21, 52–53, 58, 68–69,
70, 71, 80–82, 86, 91, 93, 95, 96, 136;
control, 37, 73. *See also* air pollution, water pollution
population: growth, 20, 35, 39, 64, 71,
115, 134; control, 38, 80–81, 91, 117,
130–31. *See also* overcrowding
The Population Bomb, 35, 38–41, 130–31
Portland, Oregon, 62–63, 97–99
progress, 13, 15, 33–34, 51, 86, 127, 134
public health, 37, 56–58, 65, 103–4; air
pollution and, 17–19, 29, 49–51;
hazardous waste and, 151–55; lead
poisoning and, 43–48
Public Health Bulletin No. 306: "Air
Pollution in Donora, PA," 16–19
public service announcements,
100–101
Puyallups, 77

racism, 61, 88, 135
radicals, 10, 44
Ramparts, 6
rats, 39, 61, 85, 88, 90, 92
Reagan, Ronald, 69, 107; Remarks
before the American Petroleum
Institute, 114–18
recreation, 5, 25, 33–34, 72, 75–76, 95,
106, 109
regulation, 9, 71, 104, 127
Rhine River, 4
River Rouge, 36, 53
Rockefeller, Nelson, 69
Roland, Robert A., 140; Statement
Regarding Superfund, 146–50
Roosevelt, Theodore, 5

Sanders, Dr. N. K., 37; "The Santa
Barbara Oil Spill: Impact on the
Environment," 54–55

Vietnam War 6, 8, 68, 85–86

Village Voice, 43–48

water pollution, 3, 36, 58, 72–73, 91,
    116–17, 119–21; Clean Water Act,
    72–73; in the Cuyahoga River
    94–95; in Detroit 52–53
Water Pollution Control Act (1971), 37,
    119–21, 139
"We Have Met the Enemy, and He is
    Us," 84
"We Will Stop the Bulldozers," 122–24
Whole Earth Catalog, 9, 36, 42
wilderness, 5, 6, 8, 14, 57, 106; preser-
    vation, 25–28
Wilderness Act (1964), 6, 14
Wilderness Society, 10, 14, 25

Young, Neil, 7

Zahniser, Howard, 14; "Wilderness
    Forever," 25–28
zero discharge, 119–21
zoning, 97–98

WEYERHAEUSER ENVIRONMENTAL BOOKS